Body of Knowledge

ONE SEMESTER OF GROSS ANATOMY,
THE GATEWAY TO BECOMING A DOCTOR

STEVE GIEGERICH

A LISA DREW BOOK
SCRIBNER
NEW YORK LONDON TORONTO SYDNEY SINGAPORE

SCRIBNER
1230 Avenue of the Americas
New York, NY 10020

SCRIBNER and design are trademarks of Macmillan Library Reference USA, Inc.,
used under license by Simon & Schuster, the publisher of this work.

Designed by Colin Joh

Manufactured in the United States of America

1 3 5 7 9 10 8 6 4 2

Library of Congress Cataloging-in-Publication Data

Giegerich, Steve.
Body of knowledge: one semester of gross anatomy,
the gateway to becoming a doctor/Steve Giegerich.
p. cm.
"A Lisa Drew Book."
1. Human dissection. 2. Medical students—New Jersey. 3. Human anatomy—
Study and teaching—New Jersey. 4. New Jersey Medical School. I. Title.
QM33.5.G544 2001
611'.0071'174932—dc21
00-068781

ISBN 0-684-86207-7

For my parents,
Raymond William
and
Lois Shriver Giegerich

To examine the causes of life, we must first have recourse to death.

    —Mary Shelley

Consider your nature, you were not made to live as beasts, but to pursue virtue and knowledge.

        —Dante

Body of
Knowledge

CHAPTER 1

———————————•———————————

At mid-morning on Tuesday, April 1, 1997, a late-model station wagon, ordinary but for the smoked glass obscuring its rear windows, turned from South Orange Avenue into the main entrance of the New Jersey Medical School parking lot. From the driveway, the car made a hard left, descending immediately down a ramp leading to a submerged loading dock.

Because the height of the dock accommodated trucks, not cars, the driver parked off to the side. Unlatching the tailgate, he slid his delivery from the vehicle. Workers in another venue might have been unsettled by what emerged from the cargo hold, but in this environment no one paid the slightest attention. Here, dead bodies were a matter of course. Every day, someone was either coming—into the embalming room operated under the supervision of the NJMS anatomy department—or going—the dock also served as the dispatch point for undertakers retrieving the deceased from the medical school's sister institution, Newark's University Hospital.

Once the collapsible gurney was removed from the station wagon, the driver clicked the stretcher into position and wheeled it up a foot ramp and through a set of automatic metal doors warning away all but authorized personnel. Inside the building, he approached a second AUTHORIZED PERSONNEL ONLY sign, where he pressed the buzzer outside a brown, windowless metal door.

Twenty seconds later, the door swung open. "Got one for you," said the driver, handing a folder to a stocky man in his early sixties with the erect posture of a military veteran. The driver, an employee of Funeral Service of New Jersey, a company dedicated to the transportation of human remains, ignored a surge of air redolent with chemicals. Pulling the gurney behind him, he entered a room cast in drab institutional yellow, its purpose distinguished by two stainless-steel cribs with drainage basins, each flanked by a pair of fifty-five-gallon chemical drums. In the corner, next to a third table, stood a fixed eighteen-inch single-blade jigsaw.

Roger Faison accepted the folder and tossed it onto a countertop. Excusing himself, Faison returned to the room a moment later with a medical school stretcher onto which he and the driver transferred the body. Plucking a receipt from the folder, Faison signed it and slipped it to the driver, who was on his way out the door within five minutes of arrival.

Faison thumbed absently through the folder and considered his schedule. It was a slow week; the students were gone, as were most of the faculty. Along with the medical researchers and support staff, Faison usually remained behind during spring break. He didn't mind staying put. In fact, he rather liked having the place more or less to himself, if for no other reason than it decreased the number of emergencies, real and imagined, requiring his immediate attention.

Through the years, Faison had used the respite to catch up on paperwork and other assorted tasks that tended to be pushed aside while, in the laboratory one floor above, the first-year students were enmeshed in the medical school initiation known as Gross and Developmental Anatomy. From January through April—desiring to acclimate the students to the rigors of medical education early, NJMS, unlike the majority of medical schools, scheduled the mandatory gross anatomy curriculum during the second semester—chaos was the order of the day, every day. The pandemonium would resume the following Monday; until then, Faison set the pace. He put off the embalming until Wednesday.

Faison wheeled the gurney into a walk-in refrigerator. One hundred eighty-five corpses, most wrapped in clear plastic bags, lay inside. All but ten rested on open-shelved compartments arranged in a grid twenty-five rows across and seven rows deep. The most recent arrivals were on gurneys, a temporary arrangement until space on the grid became available.

Embalming bodies for a medical school was not what Roger Faison had in mind when, in 1957, he applied his GI bill toward a degree in mortuary science. He had emerged from the navy a firm believer in the American dream, believing "all that crap that I read about economics in magazines about how if you work hard, the world is yours." In Faison's case, this meant becoming the owner and operator of the best funeral home in the Brooklyn neighborhood where he grew up, Bedford-Stuyvesant.

Fresh out of mortuary school, he landed an apprenticeship and began to expand himself academically. Understanding that mortuary wasn't the only science he'd need in order to make a name for himself in the funeral industry, Faison enrolled at Fordham University. Four years later, he departed with a degree in economics.

A submariner at the height of the cold war, Faison's navy stint only heightened his sense of adventure. In the navy, there had been exhilaration in spending weeks tracking Soviet submarines below ocean surfaces. In business, the thrill came in the pursuit of financial success.

Traditionally, in small towns across America and especially in the South, the black community revolved around the churches and the mortuary. The local undertaker was a professional, a man of dignity and grace, a man who, before civil rights laws prevailed to change the scope of race relations, served as the nominal link between his community and the prevailing white power structure. Brooklyn, which, in the early 1960s, still prided itself on being the biggest small town in America, was no different. With a gentle demeanor that masked a droll sense of humor, Faison settled into the niche. His business took off. Bed-Stuy brought him its dead; he, in turn, provided compassion and understanding, services he brought with him also to Manhattan's Upper West Side after borrowing the money to open a second parlor there.

Back in Brooklyn, a borough with a population larger than all but a handful of the country's biggest cities, Faison conceived a plan he hoped would appeal to the huge untapped market residing in Bed-Stuy's tenements and housing projects. Knowing a funeral could be arranged for far below the going rate, $1,000, Faison launched a marketing campaign, papering the projects with leaflets guaranteeing a complete funeral, sans burial expenses, for $500. Citing a law prohibiting such a blatant form of advertising, state regulators told him to knock it off. The competition took an even dimmer view, twice phoning bomb threats to Faison's mortuary within an hour of a scheduled funeral. Faison took the hint and reverted to the standard fee assessed by the city's other mortuaries.

Economics, the very subject he'd gone out of his way to master, proved to be his downfall. When, in the early 1980s, conglomerates began buying up New York mortuaries en masse, Faison refused to sell and gamely tried to compete. With the advantage of volume economics, the bigger companies eventually undercut the competition; in other words, the same economic theory that two decades before had brought Faison bomb threats was now turned against him. In 1986 he gave up, sold the funeral homes, paid off the banks and moved to New Jersey and a job immune to the trials imposed by the free-market system.

Embalming at NJMS brought with it a different set of predicaments. In the private sector, embalming emphasized presentation—the undertaker's objective, on behalf of the deceased, was to create a lifelike appearance engineered to last for the intervening period between death and burial. At the medical school, aesthetics went out the door; preservation became paramount. Faison no longer cared what the bodies looked like; his only concern was to ward off decomposition. The bodies in a Gross and Developmental Anatomy laboratory had to last fourteen weeks lest Faison incur the wrath of faculty and students alike.

Each year, at their own request, the donated bodies of nearly a hundred men and women were transported to the basement of the ten-story NJMS medical science building, located in the heart of Newark's Central Ward. Of that number, slightly more than half ended up in the anatomy lab: forty-five to be dissected by medical students, another twenty-five dissected by students attending the adjoining institution, the New Jersey School of Dentistry, ten to fifteen more to train surgical residents and a handful beyond that deployed to assist qualified surgeons in the development of new surgical procedures and techniques.

Bodies that didn't go to the laboratory normally wound up in the university's medical research wings. While some donors requested their remains be channeled to researchers with a specific area of expertise, most placed no restrictions. The decision about placement was usually Faison's. His was a simple formula: Bodies dispatched to the gross anatomy laboratory had to merit an "AA rating." Meaning a lean body, not prone to decay.

Preventing decomposition became all the more difficult when, two years after Faison came to NJMS, the federal Occupational Safety and Health Administration (OSHA) showed up for a routine inspection of the first-floor gross anatomy laboratory, four windowless rooms with dropped ceilings. Citing poor air quality and substandard ventilation, the agency declared the formaldehyde used to preserve the dead to be hazardous to the living and ordered the school to shut down the lab. The action, halfway through the semester, triggered a spate of appeals from the school administration to a bureaucratic entity that dictated that the lab doors remain locked until Faison came up with an alternative to formaldehyde as an embalming solution.

The government regulators, however, failed to take into account the perseverance of the students. Undeterred, they began showing up at the lab in the wee hours of the morning. Employing the hammers and chisels issued them for the purposes of dissection, they unhinged the doors and went about their task nocturnally. The school looked the other way. When the faculty, support personnel and OSHA monitors arrived the next morning, the doors were again hinged and locked.

As the students pursued their course of study by night, Faison spent his days grilling chemists, funeral home directors and medical school embalmers across the country for answers. Faison's mission—to find a low-toxicity chemical with long-term preservation qualities—tested the creativity of the best minds in chemistry and the mortuary sciences. Finally, two weeks after they first locked the doors, Faison proposed to OSHA that formaldehyde be substituted with a concentration of phenol, alcohol and glycerin. The compound smelled as bad as formaldehyde, and there were no guarantees that it would preserve a human body over fourteen weeks, but with the administration and anatomy faculty breathing down his neck, Faison had no other choice. To the relief of all, the agency accepted.

Because the cadavers in the lab were already embalmed with formaldehyde, a compromise was reached to allow the resumption of the semester: Permitting the lab to reopen, OSHA stipulated that its representatives be present for the weeks remaining in the academic term to monitor toxicity levels. For the rest of the semester, the students and faculty worked with overhead measurement booms, similar to those used by television news crews. The students resented the intrusion, and though the booms over the tables were an irritating distraction, they felt they gained a measure of retribution in the obvious discomfort that the OSHA team had with the lab's primary activity.

Despising the substitute compound, Faison never again embalmed with formaldehyde. That OSHA subsequently forced other medical schools to switch to phenol did nothing to allay his consternation. Especially when a counterpart at an Ivy League institution gloated, "You guys just don't have enough clout." Faison knew it to be true: Among the prestigious medical schools lining the Northeast Corridor from Boston to Washington, NJMS was pretty much relegated to the status of poor cousin.

Ten years after the OSHA crackdown, Faison still struggled to find an adequate level of phenol-glycerin to keep the cadavers viable. Using weight as a criterion, most of the time he guessed correctly. The X factor was metabolism. There was no way to know which bodies would most effectively metabolize the compound and which would begin to decompose the moment they were removed from the refrigeration unit.

Before OSHA, Faison took pride in his embalming. In the years after, the caliber of his work became a source of embarrassment: "When we used formaldehyde these bodies were standing tall and right. Now, halfway through the semester, I'm ashamed. Just ashamed."

Governmental interference notwithstanding, the medical school provided Faison with a secure environment. Unencumbered by obligations to lending institutions, Faison no longer had to worry about the competition. With money no longer a factor, an unexpected dividend emerged: the students. Fascinated by their swagger, confidence and how, despite some unbelievable setbacks, they always bounced back, Faison loved interacting with them. Their youthful optimism was contagious, and best of all, every year a new batch arrived—filled with just as much brio as the class before.

The downside of academia resided in office politics. Faison, with a front-row seat for the internecine squabbles both inside and outside the anatomy department, had never seen anything like it. The anatomists blamed the cell and tissue biology faculty for inadequately preparing the students for anatomy; the senior surgical fellows pointed the finger at the anatomists for not fulfilling their obligations to the students. On and on it went. Within the anatomy department every decision, be it by administrator or faculty, triggered endless second-guessing. "Sometimes it amazes me that anyone ever learns anything here," said one anatomy instructor, who added, "Against all odds, though, it happens."

Faison may have witnessed firsthand the endemic political infighting, but he also had a place to escape: a refuge seven floors below the academic battlefield on the G level, home of the NJMS Department of Anatomy.

The first thing Roger Faison did upon arriving at his sanctum early on the morning of April 2, 1997, was to remove from the refrigerator the body that had arrived the previous day. Maintained at a constant temperature of 34 degrees, the refrigerator preserved the integrity of capillaries that would otherwise collapse should the thermal reading inside the unit drop below freezing.

With the assistance of Dr. David Abkin, a former Russian surgeon who in addition to operating the anatomy lab supply room also served as Faison's assistant, the mortician shifted the body from the gurney onto one of the stainless-steel basins. While the refrigerator was kept warm enough to prevent the deterioration of veins and arteries, its algidity nonetheless dictated a ten-hour thawing before the onset of embalming.

At four o'clock that afternoon, Faison slipped a Miles Davis CD into his boom box and returned to the body. Peeling back the sheet covering the head, the mortician made a three-inch incision in the right side of the neck. From the top of the chemical barrel, he retrieved and fit hypodermic needles onto the ends of two rubber tubes. He inserted one needle into the right common carotid artery, the other into the right internal jugular vein. Faison flicked a switch and, above the table, a pump began to rumble, the fifty-five-gallon drum started to gulp. The mortician watched as the phenol-glycerin trickled through the tube attached to the common carotid artery, causing blood to simultaneously drain from the second tube into a barrel marked for medical waste.

Ninety minutes later, when the fluid emerging from the drainage tube ran clear, devoid of blood, Faison switched off the pump. Removing the needles from the neck and then the tubing, he inserted them in a red medical-waste box, and he returned to the body, where he tied shut the incision with heavy-gauge embalmer's thread.

Faison moved to the foot of the body, removed the toe tag and checked the name on the tag against the paperwork brought by the funeral service driver: Lewis. He tossed aside the tag and retrieved a translucent orange bracelet from a box on the counter. A number had been written on the bracelet with a Magic Marker: 3426. Faison scribbled the number across the front of the folder and fastened the bracelet around the cadaver's left ankle.

The mortician flipped through the paperwork and began dividing it into two piles. Most of it—the death certificate, information about notifying the next of kin once the ashes became available—he earmarked for Essie Feldman's office on G level. For his own files, Faison copied a few documents and placed them in an office cabinet. Faison scanned the obituary tucked into the folder and learned the man he'd just embalmed had been an educator. A teacher, Faison thought, very appropriate. He placed the obit back in the folder; on a copy of the receipt left by the funeral service, he noted: "Good subject. Possible Medical Gross."

Abkin helped Faison place the body in a large, clear plastic bag. They transferred Number 3426 to the gurney, which Faison then wheeled back into the refrigerator, officially designated on the building registrar as Room A526B.

On the wall above the clutter of folders, memorandums, written requests for donor forms, half-finished letters, partially completed syllabuses and other academic flotsam littering Essie Feldman's work area hung an enlarged photocopy declaring: "A Clean Desk Is a Sign of a Sick Mind." That being the case, Feldman was the picture of mental health.

The morning following the embalming, Faison deposited Number

3426's folder atop the pile and spent a few moments with Feldman catching up on the latest gossip. All anatomy department scuttlebutt passed through Feldman, whose ebullience—everyone, save the department chairman, Dr. John H. Siegel, received the same salutation, "honey"—penetrated the gravitas that permeated the halls of medical education. Flighty in personality and stylish in dress, Feldman stood out in a recondite atmosphere where grave demeanors were as common as white lab coats. Given her personality, Essie Feldman would, on the surface, seem an unlikely liaison between the medical school and the general public when it came to a topic most people are loath to discuss: death and its aftermath.

Feldman hadn't been predisposed to counsel the living about that which would transpire after they died; instead, the position chose her. In 1969, Feldman first reported for duty at the medical school in the capacity of secretary, and to the payroll department, a secretary she would remain during a tenure that would eventually exceed thirty years.

On Essie Feldman's first day on the job, the New Jersey College of Medicine, as it was then known, had been operational barely two years. The school was the offspring of the Seton Hall College of Medicine and Dentistry, which, in 1956, had the distinction of being New Jersey's first medical school. Several reasons have been advanced as to why New Jersey didn't have a school of medicine until the last half of the twentieth century, the most popular being that it took that long for the state's legislators to reject antivivisection statutes imposed by their Puritan political forebears. For whatever reason, New Jersey politicians believed laws designed to prevent the mutilation of animals applied also to the dissection of human cadavers.

Academically and also by the real barometer used to evaluate the success of medical schools—securing federal grant money—Seton Hall was an unqualified success. Its Jersey City campus, perched on the Hudson River overlooking the Manhattan skyline, proved to be a magnet for leading researchers, many of whom abandoned the epicenter of U.S. medical research—Boston—to join the fledgling school. Unfortunately, the school's bottom line did not match its achievements in the classroom and laboratory. Start-up costs far exceeded expectations, and even the federal grants couldn't stave off the fiscal reality of operating a medical school. It didn't take long for the Seton Hall College of Medicine and Dentistry, under the domain of the Catholic university in South Orange bearing the same name, to realize its books would never balance. By the early 1960s, the school started casting for a benefactor to bail it out; five years later Seton Hall's knight arrived in the form of the State of New Jersey.

The team undertaking the state's first order of business, finding a site to relocate the campus, settled on a 150-acre estate in yet-to-be-suburbanized Morris County. Located in the northwest quadrant of the state, the property offered everything but an ethnically diverse employment pool. Exploiting Morris County's shortcoming, Newark's Democratic mayor Hugh Addonizio presented the Democratically controlled state legislature with an alternative: a city with an existing ethnic workforce to staff a major medical school and hospital complex and, for the campus's construction, contracts that would provide jobs for thousands of laborers represented by the unions that greased the wheels of New Jersey Democratic machine politics. To make room for the medical school and hospital, Addonizio promised to condemn 150 acres in the Central Ward, just west of downtown.

The state legislature may have embraced Addonizio's plan, but not so the one thousand men, women and children residing within the boundaries of those six square impoverished and crime-infested city blocks. They stood to lose their homes. In July of 1967, as Addonizio's recommendation moved closer to reality, their anger, aggravated by multiple other factors, exploded into three nights of mayhem that left twenty-six dead.

Two months later, a medical school that didn't want to be there welcomed its first class of students to a neighborhood that didn't want them. Many years passed before the Central Ward accepted the school and medical center, and despite outreach programs that in 1994 earned it an Outstanding Community Service Award from the American Association of Medical Colleges, to this day the resentment among some residents continues to run so deep that they profess to prefer dying in an ambulance en route to Beth Israel Hospital, five miles away, than accept treatment at the UMDNJ-University Hospital in their own backyard.

The relocation to Newark cost the medical school dearly. Much of the faculty abandoned the institution, and with them went federal research funds and burgeoning prestige.

Among those making the move from Jersey City to Newark was a young and theatrical anatomy instructor named Anthony Boccabella. A self-described "hotshot," Boccabella tended to pepper his lectures with double entendres. Given the subject, the possibilities were endless. He was no less flamboyant outside the classroom, boasting of being the first medical student ever to arrive at the University of Iowa piloting his own plane. Later, while serving in the joint capacity of administrator and instructor at NJMS, he obtained a law degree, a pursuit he credited to being made to feel like a black sheep in a family of lawyers.

In 1971, a year after the New Jersey Medical School was folded into the statewide system of health education facilities now known as the University of Medicine and Dentistry of New Jersey, Boccabella became the chairman of the anatomy department. He would have preferred to focus on his field of expertise, endocrinology, the study of the thyroid, pituitary and other endocrine glands. Still, he appreciated the value of anatomy as the basic foundation of medical school and medical practice. "The rest is just dressing," he fondly pointed out.

During its brief tenure, Seton Hall suffered no shortage of cadavers. Boccabella recalled that most "were not there voluntarily," a polite way of saying the bodies had been secured from morgues. When the state uprooted the Jersey City facility, the corpses were brought to Newark along with desks, examining tables and other amenities crucial to the teaching of medicine.

The surplus no longer existed by the time Boccabella took over the program. Then, as now, four to one is the accepted ratio of medical students to cadavers in domestic human anatomy labs, and rarely in U.S. schools are more than four students assigned to a table. At some European schools, the students aren't quite as fortunate, with eight or more assigned to a single table. More unfavorable are the conditions in European countries that lack an adequate system for securing bodies, a situation which dictates that instructors dissect while students are relegated to the role of spectators. In the United States, Boccabella observed, students get "hands-on experience. They're able to play with it, cut through it and learn from the body. We're very, very lucky in this country."

In 1971, the dearth of cadavers in Newark nearly forced Boccabella to adopt the European custom. At each table, six medical students and one dental student worked on a single cadaver. The dental students dissected the head; the medical students studied the rest of the body. With the situation reaching the "crisis" stage, Boccabella undertook a desperate move.

Inspired by medical science's advances in the field of transplantation, Americans were just becoming aware of organ donations. As a scientist, Boccabella knew of an active registry in Washington listing the names of men and women who wished to donate, postmortem, their corneas. Playing a hunch, Boccabella contacted the registry. If these people were willing to pledge their corneas, he reasoned, perhaps they'd be amenable to contributing their entire bodies. A "gracious and unthinking" clerk provided Boccabella with precisely what he desired: the names of ten thousand New Jerseyans on the registry. Before the year was out, all ten thousand received from Boccabella a "passionate" letter.

Simultaneous to dispatching the letter, Boccabella began making the rounds of the state's burial societies, organizations of senior citizens banded together to minimize the high cost of death. He started his speech with an appeal to altruism, calling the donation of a body a gift in perpetuity, an unselfish act of ultimate benefit to every man, woman and child healed by the physician who profited from its myriad lessons. Boccabella assured his audience that in the lab they would be granted complete anonymity. Neither faculty nor students would know their identities. Boccabella followed the entreaty with an offer he hoped they'd find difficult to refuse: a postdissection cremation paid in full by the State of New Jersey. After cremation, he promised the ashes would be returned to the families for disposal as they saw fit.

From Sussex County, in the farthest corner of the state, to Cape May, New Jersey's southernmost point, Boccabella delivered his spiel. On the power of personality, he soon had people discussing openly a subject previously avoided or dealt with surreptitiously.

His name became a euphemism, a punch line: A husband and wife who jointly willed their bodies to NJMS used to joke it was time to "call Dr. Boccabella" whenever one of them was laid low by a cold or other assorted minor maladies.

As often as not, Boccabella could count on at least two or three burial society members to bring an attractive next of kin to a meeting. Recently divorced and constantly on the make, Boccabella sought them out after his presentation to inquire if they'd enjoyed his discourse. Those responding in the affirmative received a donor application; those who filled it out on the spot risked subjecting themselves to the doctor sidling up and whispering: "I already own your body and I promise to take care of it, so why wait?"

One of the worst pickup lines in history notwithstanding, Boccabella's sales pitch had the desired effect. The donor application forms that began to trickle in soon reached flood level. Unable to handle the flow, the department chair turned to his secretary.

Essie Feldman, with no training in medicine and no experience dealing with the imminence of death, received from Boccabella cursory instructions on how to respond to prospective donors. Feldman turned out to be such a natural that, before long, she was in charge of the cadaver procurement program.

The job required a gentle touch and Essie Feldman had it. She became expert at the art of reassurance, especially, as was often the case, when a donor had second thoughts about his or her decision. In most cases it was a loved one who gave donors pause, not so much spouses—a surprising number of husbands and wives submitted donor forms in tandem—as sons and daughters. Children seemed to have the most difficulty wrapping their minds around the idea of what became of a human body donated to science. Sometimes Feldman spoke directly with the conflicted party and explained, as best she could, the program's importance. Usually, the appeal worked; when it didn't, Feldman cast no aspersions. It's important, she'd tell those withdrawing their names from the donor file, that everyone in the family be comfortable with the decision.

Although contact between Feldman and donors was limited almost entirely to mail and telephone, a personal relationship sometimes developed between a benefactor and the woman to whom he or she had entrusted his or her body. They'd relay news of weddings and births and anniversaries. One man sent vacation snapshots from ports of call around the world. "Maybe I'll see you soon, Essie!" he'd scribble jokingly across the back of the photos. After he died, his family invited Feldman to his memorial service. She attended.

Essie Feldman's job also brought with it a fair amount of sadness. The task she performed was the result of a decision predicated on an issue troublesome to contemplate—human beings tend to shy away from considerations of what lies beyond this mortal coil—and even more difficult to articulate. Still, after a time, Feldman noticed a trend: Most of the donors were college-educated, and according to handwritten addenda or notations on the forms, the majority also participated actively in mainstream religions.

Occasionally, she'd speak with a man or woman who expressed the wish that by making a donation someone else would be spared some level of pain and suffering. One afternoon she took a call from a woman who, in a halting voice, detailed a twenty-year battle with schizophrenia. "Have you any interest in examining a broken mind?" she inquired. Feldman assured the woman—in her late thirties, she was far younger than the people who usually contacted Feldman— that her gift would be routed to scientists conducting neurological research. The woman began to weep and then regained her composure. "Thank you, thank you," she said quietly. "I don't want what happened to me to happen to anyone else."

Of course, Feldman's life was not always the stuff of high drama. Most of the time, her workload filled the description ascribed to the job by the payroll department. In the eyes of the State of New Jersey, which paid her salary, Essie Feldman was a secretary. Meaning a good portion of each day was devoted to routinely filing paperwork submitted by men and women who'd made a remarkable and difficult decision.

Such was the case on April 3, 1997, when Roger Faison walked into her office and plopped Number 3426's folder on the ever-present stack of papers atop her desk. Later that day, after Feldman entered on the filing tab the name of the deceased and the number assigned to the cadaver, she dutifully placed the folder in a file cabinet. There, Number 3426's dossier would remain until Feldman received word from Faison that the body, by either advancing medical science through research or serving as an anatomical blueprint for a quartet of medical students, was being dispatched to the crematorium.

Upon the return of the remains to the medical school, Feldman would swing into action again, contacting the family, informing them to soon expect a package the size of a shoe box. On average, two years passed between the time NJMS received a body and the follow-up letter from Essie Feldman. A lot can happen in two years: Next of kin occasionally move without providing a forwarding address; older relatives often join their loved ones in death. Stubbornly determined, Feldman went to great lengths in the effort to find someone, anyone, to properly disperse the ashes, in one instance devoting six years to the search for a sole, surviving son. She finally located him in the Midwest.

The ashes of the men and women whose relatives eluded Feldman's dragnet were stacked four high on a shelved rack in a hallway outside Faison's office in the basement. When the number of boxes approached two hundred, Faison transported them to a Linden cemetery for interment in a special plot overseen by the medical school. For years, Faison had urged the administration to commemorate the site with a small marker acknowledging the gift of those interred there, and finally, in 1999, the school accommodated him.

For eighty-nine weeks, Number 3426 remained in the refrigeration unit. In early December 1998, Faison started to review his files, setting aside those marked "Possible Medical Gross." In choosing bodies for Medical Gross and Developmental Anatomy, Faison knew what he was looking for: the elusive "AA" ratings. And Number 3426 fit the bill.

Six days before Christmas, Faison and Abkin removed Number 3426 from the refrigerator and placed the body on a stainless-steel tray measuring thirty inches by seventy-five inches, aerated at the bottom by twenty-eight drainage holes. Placing the tray on a gurney, Faison and Abkin covered Number 3426 from head to toe with five frayed towels and wheeled the gurney onto a freight elevator located just outside the entrance to the embalming room. When the gurneys in the elevator numbered three, Faison turned his key and pushed the button marked B. At the B level, the doors opened into the supply room attached to the gross anatomy laboratory serving the students of the New Jersey Medical School.

The NJMS laboratory is actually four interconnected laboratories, lettered A, B, C, D, each with eleven dissection tables. Lab A is tucked off by itself. The other three form an inverted F. Along the spine of the F are light boxes, a place where the study of the dead is supplemented by photographic representations of the living: X rays, magnetic resonance images (MRIs) and CAT scans.

With a dropped ceiling, fluorescent lighting and no windows, the gross anatomy lab is mundane, sterile. No art adorns the walls. Only the orange storage cabinets and the stained orange and yellow cushions on the rolling chairs clumped around the dissection tables break the monotony of the room's dominant color, off-white.

The tables, $3,800 new, stand forty-two inches high when the hoods are pulled back and secured below the table. When the hood is closed, it raises the height of the table to sixty inches. Anyone looking at the five-sided beveled hood would know precisely what lies inside.

By the afternoon of December 19, Faison and Abkin were half finished with their task. With the tables in Labs A and B already occupied, they began filling Lab C. Two at a time, they took the three gurneys from the elevator through Lab B into Lab C. Number 3426 was wheeled by Faison to the last table in the first row of the room.

Joined by Abkin, Faison counted to three and the two men lifted the tray onto Table 26. Wordlessly, Faison shut the hood and began pushing the gurney back to the elevator. Eighteen AA-rated bodies remained in the basement. Roger Faison checked his watch and did a quick calculation: Six more trips, and then, finally, he could begin his Christmas break.

CHAPTER 2

──────────●──────────

I f twenty-two years spent in operating theaters and emergency
rooms had imparted no other lesson, it was the recognition that
certain courtesies must be extended. So, in lieu of doing what she
really wanted—popping open the hood of examination Table 26—
Sherry Ikalowych made nice.

She hadn't planned on discussing her children with her new lab
partners, but when Jennifer Hannum raised the question, Sherry
automatically recited names, ages and the most recent academic and
athletic accomplishments. Jen politely acknowledged the information.
For Jen, with a minimum of seven years of medical education on the
horizon, kids were not yet a priority. Her short-term goal, vis-à-vis
any type of future partnership, was to perhaps go on a date before the
end of the semester, preferably with someone other than another
medical student.

The last time Jen had given parenthood even a modicum of thought
was during a childbirth episode on The Learning Channel. Jen, a TLC
junkie addicted to the cable outlet's real-life medical programs, provided details of the show's portrayal of a mother's journey from labor
through birth. The description drew the attention of Udele Tagoe,
who was leaning toward a career as an ob-gyn.

"How were your deliveries?" she asked Sherry.

In full-parent mode, Sherry provided the details, explaining that
the first two were natural, the last a C-section. Twins, she added.

All business, Udele turned her attention to the bone box, a wooden
satchel the size of an overnight bag that contained an improbable
number of human skeletal remains culled from cadavers used by the
university's research wing. Ostensibly, the only activity required of
the 181 members of the Class of 2002 on this, the first day of the second semester of their first year of medical school, was to inventory

the contents of the box issued to each of the forty-five gross anatomy lab's dissection tables.

Beyond that, the underlying purpose of the lab was an opportunity for the four lab partners assigned to each table to get acquainted. At one time, students were allowed to choose the person or persons with whom they wished to share the four grueling months spent in the anatomy lab. Familiarity, the anatomy department found out, minimized learning potential. The group dynamic, which demands that someone take the lead, had a way of breaking down when friends deferred to one another. When no one led, no one learned. Out went the buddy system.

In its place, the department teamed lab partners based on observations made by instructors who had encountered the students during the first semester. Overachievers were placed with students requiring a shot of intellectual adrenaline. Academic matchmaking being an inexact science, the system remained far from perfect: Sometimes it worked; just as often it didn't.

In addition to taking inventory of the contents of the bone box while getting to know their lab partners, students so inclined could also use the first day of lab to meet, in a manner of speaking, the person on the table, an individual who in life made the decision to impart, in death, the very foundation of medical education.

Having waited a long time for this day to come, Sherry was so inclined. She wanted to see who lay on the gurney inside the table tucked into the left corner of Lab C. Before that could happen, Sherry decreed, a matter of medical protocol, not to mention common courtesy, needed resolution. "We can't open it until everyone is here," she declared. Everyone being Ivan Gonzalez, the fourth member of the team designated to dissect at Table 26. Until Ivan's arrival, the rest of the team agreed to occupy themselves with the bone box.

"Metacarpals? What are those?" Jen asked, reading from the checklist.

Sherry plucked from the box a skeleton of a hand, the joints as well as the multitude of tiny bones in the wrist held together by surgical staples. She pointed to the bones below the first knuckle. "Metacarpals," she stated matter-of-factly.

"Oh, yeah. I didn't prestudy," Jen confessed, her face turning slightly crimson.

She brushed a stray hair out of her eye. As a concession to the insidious first-year smell—the by-product of embalming fluid combined with the odor given off by decomposing bodies—Jen had debated whether to crop the auburn hair that reached past her shoulders, framing an angular face. Fearing there would be enough demands on her time during the coming semester without having to shampoo every day, Jen nearly went ahead and made the appointment with the hairdresser. In the end, she couldn't do it. The hair remained long; a ponytail became the concession.

Sherry, who already wore her hair short, showed up for the first day of the second semester in a color four shades lighter than it had been on the last day of the first semester. The blond highlights showed off the defining facial characteristics: blue-gray eyes, which, depending on the circumstances, either hid or betrayed her every emotion.

Jen glanced again toward the brown metal door leading directly into Lab C. She fretted that Ivan might be stuck in his hometown of Miami or, in the event he'd made it back from Florida, had decided to spend the day at his apartment. There were flurries in the forecast, and, except for a trip to California, Ivan had never ventured north of Georgia when he'd arrived in Newark to start classes the previous August. Not only had he never driven in snow, he'd never even seen the stuff.

Sherry Ikalowych didn't know this about Ivan Gonzalez. To Sherry, Ivan Gonzalez was a name and, vaguely, a face. The name had been one of four printed under the heading "Table 26" on the gross anatomy lab assignment list posted by the anatomy department prior to Christmas break. The face may or may not have been that of a guy with a wispy dark mustache and mischievous brown eyes who'd been stationed at the opposite end of Sherry's cell and tissue biology lab the previous semester. Whoever this Ivan Gonzalez might be and whatever he looked like, Sherry wished he'd hurry it up. She was preternaturally curious, and not opening the hood on Table 26 went contrary to her every instinct.

"I'm betting it's a man," Sherry said to Jen. Though they'd also been in the same cell and tissue biology lab the previous semester, Jen and Sherry had mostly been nodding acquaintances. Jen may not have known Sherry, but she certainly knew of her. Everyone did. In a sea of twenty-somethings, Sherry stood out. At forty, she was the oldest student in the NJMS Class of 2002, and furthermore, as a nurse-anesthetist with a master's degree in nursing and a curriculum vitae that included an assistant professorship at Columbia, Sherry possessed, hands down, the class's most impressive résumé.

At NJMS, nontraditional students—those who took something other than a direct route from an undergraduate program to medical school—were the rule rather than the exception. Sherry, in fact, wasn't even the oldest student matriculating through the New Jersey Medical School. That distinction belonged to a member of the Class of 2001 who passed his fiftieth birthday shortly after he began classes. Still, it was her four kids that really set Sherry apart.

To her classmates, some barely six years older than Sherry's oldest son, Jason, fourteen, managing medical school while attending to the needs of four children was unfathomable.

Those attempting to imagine the constraints on Sherry's time needed only recall the first-semester blitz of biochemistry, histology (cell and tissue biology) and genetics. According to the wisdom passed on to them by upperclassmen, the first-years expected the fall term's academic diet to be comparatively easy in contrast to the hell of gross anatomy and physiology, which lay ahead in the second semester. Accepting this assessment at face value, some members of the Class of 2002 came to NJMS expecting to coast through first term. Medical school, they discovered, has a way of stopping academic inertia in its tracks. In fact, the initiation proved so arduous and time-consuming that, of necessity, many cut by half the hallmark partying of their undergrad years. Cognizant of how profoundly the study load had altered their own lifestyles, no one under the age of twenty-five knew what to make of Sherry. Even the most immature students were astute enough to realize that, unlike the weekly beer bash, parenting was not an option subject to elimination.

The secret to Sherry's juggling act, a confidence she shared with anyone who asked (and many did), was twofold. Her husband, Jerome, a detective-sergeant with the Closter Police Department in the small Bergen County community where they made their home, picked up much of the slack. As for housekeeping, washing and meal preparation, Jerome and Sherry splurged on a live-in nanny who could be there when the kids—Jason, eight-year-old Sarah and the six-year-old twins, Stephen and Paul—got home from school.

Nonetheless, Sherry was not averse to reality: The kids often demanded more than a dad or a nanny. Adjusting her schedule accordingly, the quotidian began with the alarm sounding at 5 A.M., an hour that provided sixty minutes of study time before the kids arose at 6. After breakfast, Sherry began the hour-long drive to Newark to be at the medical school's George F. Smith Library of Health Sciences by 7:30. Another ninety minutes of studying and it was time for up to six straight hours of lectures and labs, followed by more studying. Arriving home around 6 P.M., she'd have dinner with Jerome and the kids and then a couple hours of family time. At 8:30, she trundled the children off to bed so, by nine o'clock, their mother could end the day as it had begun: hitting the books. Rare were the nights that Sherry managed to keep her eyes open past eleven o'clock. Six hours later, the routine began anew.

Unfortunately, four children had a way of constantly fouling the planned order of each day. The twins, by dint of being the youngest, demanded constant attention and Sarah, the only girl, was at an age that required unending reassurance from the only other female relative in the house.

Usually it was Jason who got short shrift. Beyond Sherry's mandatory presence at Jason's wrestling matches and whatever conversations they could manage between school and the trip home when she picked him up after practice, Sherry didn't have a lot of time left for her eldest. Jason had been a teenager for only two years, but he was already well versed in how to play the guilt card.

Near the conclusion of the previous semester, on an evening when she absolutely had to study for a genetics exam, Jason had badgered her incessantly to engage him in a game of Scrabble. Capitulating, Sherry caved in, transferring her remorse to the unopened backpack bulging with the textbooks and lecture notes critical to succeeding in a class that, thus far, had caused her no end of consternation.

The next morning and all the next day, Sherry played catch-up. Relying on guile and the reflected knowledge that came from hanging around physicians for twenty years, she passed the exam and, ultimately, the course.

Jen Hannum had been immensely pleased when she saw her name listed with Sherry's for assignment to Table 26. Intending to take full advantage of that reflected knowledge, Jen recognized immediately that Sherry's presence endowed Table 26 with a level of professional competence absent from most of the other tables.

Ever competitive—"I bet on everything. I even bet on Super Bowl commercials"—Jen lightheartedly accepted Sherry's wager on the gender of the cadaver. Absorbed in the bone box, Udele remained mute.

"Don't lose those bones." Udele looked up slowly. Her brown eyes, set under exotically arched eyebrows, flashed. When involved in a project, even one as mundane as inventorying bones, she didn't like intrusions. Roger Faison, the mortician, stood there grinning. "If you lose the bones, you have to replace them. And if you don't have any money, the price is very, very high." Faison paused for effect and made a sweeping motion with his arm. "As a matter of fact, some of these cadavers are former students."

A smile crossed Udele's face, her eyes softened, the ardor evaporated. Today was not about intensity; from personal experience, Udele knew there would be enough of that later.

Six months before, on the day Udele had reported for the six-week gross anatomy summer preparatory program, there had been another bone box, another inventory and another hooded table, its content, gender-specific, a mystery. First Students, the curriculum's formal name, was aimed at boosting the academic performances of students determined by the school to be in need of a head start before the formal beginning of classes in August. Based on her academic record, Udele certainly didn't need a head start.

She had come out of Duke University the previous May with a 3.25 grade point average. Her firsthand experience with the human body limited to having once observed spinal surgery placed her among the majority of students beginning their medical education at NJMS in 1998. With the exception of Sherry, who had witnessed thousands of surgeries, the Class of 2002's exposure to anatomy, any kind of anatomy, had generally been restricted to the undergraduate dissection of a pig. A very small pig, at that.

It wasn't only Duke that captured the attention of the admissions office but rather Udele's permanent address: Muhammad Ali Avenue, a few blocks from the NJMS campus. A kid from Newark's Central Ward attending NJMS wasn't considered unusual, it was practically unprecedented. Boccabella, who had been around for over forty years, could not recall teaching a single student from the surrounding neighborhood.

It mattered not to Udele how she gained admission to the First program, only that she had the opportunity to jump-start her medical education, even if it meant she would twice—once in the summer and again during the mandatory winter-term anatomy course—be called upon to assist in the dissection of a human body. As the summer curriculum approached, Udele was unsure as to how she felt about cutting into the cadaver. Her visit to the operating theater had not left a favorable impression; later, her most vivid memory had been that of the stench given off by the laser used to singe excess fat off the patient's back.

Nor had Udele's first real experience with mortality emboldened her for what she faced during the summer. Death had brushed Udele only once, in high school, when her grandmother died. Through no fault of her own, she had been separated from the grief, detached from the mourning.

Unable to afford the airfare so that Udele and her brother and sister might accompany them to the West African nation of Ghana for the funeral, Alfred and Christophina Tagoe made the journey by themselves. Although they were short the necessary funds to finance transcontinental airfare for five, Alfred and Chris never skimped when it came to imparting to their children the importance of knowing where, and what, they came from. Upon their return, they gathered the children around the television to watch a video recording of the rituals of death as observed in their native land. The images were grainy, the effect surreal, the woman on the videotape didn't look at all like the grandmother Udele remembered from visits to Ghana. The tape distanced Udele from death, rendering it intangible, undefined.

Then, in the summer of 1998—right in the middle of the gross anatomy preparatory—blaring sirens woke Udele from a sound sleep. In and of itself, waking to the sound of sirens was not unusual; as in many urban areas, sirens piercing the Newark night were mere background noise. These sirens, however, were outside the gate of the Tagoes' town house; pulling on a robe, Udele hurried downstairs.

Officially, the incident wound up in the carjacking/stolen car file of a city once known as the nation's Stolen Car Capital. That morning's episode began on the other end of town with a carjacking and ended with the stolen vehicle striking Christophina's car, parked on the street directly in front of the gate leading to the family home. Prior to impact, the young man occupying the passenger side calculated a last-second leap from the speeding vehicle to be his best chance to avoid injury and apprehension. He calculated wrong. The cops and Udele arrived on the scene simultaneously: Christophina's car was totaled, as was the passenger. Udele took one look at the body splayed vertically across the gutter, head split open, brains splattered over the curb. Nauseated, she turned away and hurried back into the house to warn the rest of the family not to go outside.

That afternoon, Udele returned to the gross anatomy lab. The cadaver on her table was slight of build and had died at a very old age. On the first day of lab Udele had withdrawn emotionally, helped immensely by a cadaver she thought resembled a facsimile of a human being. The dissection—limited to three primary areas by the six-week window—provoked nothing even approximating anguish. Not once during the methodical dissection of the woman's thoracic cavity, arm and upper back did Udele experience the horror she'd felt in the early hours of the morning when she stepped out her door. She rationalized the emotional discrepancy between the two events by remembering that the woman lying before her had made a conscious decision to one day place her body in the care of medical students. And while the carjacker had also made a conscious decision—albeit a bad one—the difference was that between a life lived and a life snuffed out. Growing up in Newark, Udele had been around enough of the latter to clearly understand and appreciate the disparity.

Another factor had also worked to neutralize the trauma of dissection. Udele Tagoe, quite simply, loved to cut.

She liked the feel of the blade as it sliced cleanly through the skin and the tug that came when, deep inside the tissue, the forceps snagged a structure that would soon reveal itself to be a nerve or artery or vein. Every hour of every day in lab brought another epiphany. She became unsettled only once: As she assisted her lab partners in turning the cadaver over for the dissection of the back, Udele became entangled with the cadaver's arm. Getting hugged by a dead person was something Udele hadn't considered when she applied to medical school.

Inventorying the contents of the bone box upon her return to lab six months later, Udele again felt a familiar edge. She didn't speak up when Sherry wondered aloud who would make the first cut. But Udele certainly knew the answer. Without hesitation, she would do it.

The way things were going, it appeared that the first cut would occur on the same day that Udele and her lab partners got their first look at the cadaver laying on Table 26. For Ivan was nowhere in sight. Losing patience, Sherry tugged halfheartedly at the handle on the hood. It didn't budge. "Must be locked," she muttered, failing to notice the obvious: The table lacked any type of locking mechanism.

Jen kept a vigil at the door. While she didn't know him well, Jen was familiar enough with Ivan to understand he masked intelligence and a single-minded determination behind a veneer of amiable apathy. The previous semester, Jen had witnessed on the basketball court that Ivan did care, so much so that it made no sense for him to be nearly forty-five minutes late for an hour-long lab session.

Ivan had at first been wary when Jen answered the summons seeking first-year students interested in playing intramural basketball, a trepidation that evaporated when Jen became the secret weapon for ATP/2002 (named for Adenosine Tri-Phosphate, the building block protein for human energy). At the outset other teams treated Jen, the only woman in the league, as a novelty and tended to leave her alone on the perimeter, a poor tactical decision considering that five years before she'd been the point guard on a nationally ranked high school basketball team. Jen knew precisely what to do when the opposition took her for granted, and by Christmas break—not even halfway through the season—the number of three-point shots she drained from above the top of the key left her seventeen points shy of establishing a new intramural league scoring record for women.

Sensing Sherry's growing agitation, at that particular moment all Jen wanted to drain was Ivan. From checking the yellow semester schedule card, the docket that dictated day by day, hour by hour and sometimes, it seemed, minute by minute every aspect of academic life, he surely knew that lab had started at 9:30.

Udele finished inventorying the bone box. Except for the skull, femur and other bones too large to fit, all the relevant skeletal parts were present and accounted for. Udele headed for the supply room to turn in the checklist; should Table 26 ever require the larger bones, the supply room was the place to go.

At one point or another, everyone involved in the gross anatomy program at NJMS passes through the supply room, the domain of David Abkin, the expatriate Russian physician who lacked the requisite training and certification to obtain a medical license in the United States. The federal government's declaration that he was unqualified didn't prevent Abkin from dispensing homilies about medicine in general, and the Russian way of practicing medicine in particular, to the continuous stream of students who presented themselves at the supply room's Dutch door for scalpel blades, latex gloves, anatomical models, saws and chisels. In addition to the necessities, the student departing the supply room also walked away with a basic appreciation of the relative ease of U.S. medical training in comparison with the rigors of a Russian medical education during the cold war.

Abkin's pride and joy was the bone collection meticulously filed in cardboard boxes lining three metal shelves at the rear of the supply room. Before the Russian's arrival, the bones had been tossed haphazardly throughout the room, the femurs with the ankles, the wrists with the vertebras. It was chaotic, everything mixed together like that. Abkin sorted the bones, placed them in clearly marked boxes and arranged the boxes on the shelf accordingly. For students and staff who might not appreciate his organizational skills, Abkin placed hand-lettered signs on each of the three shelves. PUT BONES BACK IN THEIR PROPER BAG, the signs warned.

At the Dutch door, Udele handed Abkin the inventory of Table 26's bone box and he added it to the pile on his desk.

En route to and from the supply room, Udele noticed that several students in the other labs had decided to open their tables. In Lab C, David Murphy, at Table 25, was about to follow suit. With his own lab partners as well as those at Table 26 serving as an audience, he tugged gently on the handle of the hood; it swayed open effortlessly.

"Guess they're not locked," said Sherry. As she stared at the towels covering the cadaver from head to toe on the neighboring table, curiosity trumped protocol. She turned back to Table 26 and yanked the handle with authority; she locked her half of the hood on the bottom of the table as Udele did the same with the other side. Jen let out a little gasp, partly at the sight of the shrouded body, mostly at the vision of the dark-haired young man with the mischievous grin making his way toward the back of the lab.

Ivan Gonzalez's reason for being nearly an hour late was far from exotic: He hadn't been delayed in Miami and he hadn't listened to the weather forecast; he'd overslept. Hoping to make amends, he first apologized to Udele and Sherry before introducing himself.

"Smells great in here," he cracked. Anticipating the smell—"like something left in your refrigerator too long" is how one med student put it—Ivan had changed from his street clothes into powder-blue scrubs and white lab coat. He needn't have bothered.

"Well, it's there," Sherry said. She contemplated the towel covering the cadaver's head. After a lifetime spent witnessing life being wrested from death, Sherry understood more than the others the symbolism of the moment. Far from being a simple reflexive act, removing the towel marked a beginning; an unknown was about to become the known. Up to that juncture, the medical education of everyone at the table but Sherry had been in the abstract, as hypothetical as the endless formulas scribbled on the overhead projector during bio-chem lectures. What they were about to undertake was different. The power to defeat illness, to heal suffering, to conquer death. That is what brought the four lab partners to this building, this room and this table, and this was where it would begin.

Lost in thought, Sherry didn't notice Nagaswami Vasan material- ize, ghostlike, at the front of the lab. The course coordinator clapped his hands. "OK, everybody. That's it," he said.

Sherry checked the clock and cursed under her breath. The hour was up; the introductory lecture for gross anatomy, by the depart- ment chairman, Dr. John H. Siegel, was about to begin in the main lec- ture hall.

Vasan again clapped his hands. "Close 'em up, let's go," he ordered.

Before closing the hood, Sherry took one last look at the shrouded cadaver and gestured toward two small lumps swelling the towels covering the torso.

"Breasts," she said with an air of certainty.

"You sure?" Jen asked. With her left hand, she gestured toward the tuft of gray poking out from under the towel at the top of the table. "Good head of hair," she marveled.

"Think it's a man?" Jen asked.

"I think I see breasts," Sherry said. She shrugged. "But you can never tell."

CHAPTER 3

———————•———————

Name an ailment, and at one juncture or another Thomas Lewis had come down with it. "If it's out there, it's going to find him," his wife of forty-two years, Connie Lewis, used to say. Open-heart surgery twice, the removal of a kidney, skin cancer, ulcers.

Given all that, the fevers and chills that struck over the Palm Sunday weekend were a blip. Not one to take chances, on Monday, Tom dutifully visited his physician, who diagnosed flu and ordered bed rest and fluids. The patient took the advice by half. It was Holy Week and nothing could keep Thomas R. Lewis* in bed during Holy Week.

Tuesday morning, Connie and Tom went to confession at the parish they'd adopted twelve years earlier, following the move from North Jersey to the Jersey Shore. To conserve energy and hasten his recovery, Tom followed the doctor's orders by spending Wednesday in bed. Feeling no better, he and Connie skipped Mass on Holy Thursday. That night, at dinner, Tom sought atonement for his absence by orchestrating a variation of the Last Supper. Pertinent scriptures were read, a grandson performed the ritual of washing everyone's feet and the appropriate prayers were offered.

Devotion to God and the Roman Catholic Church were nothing new for Tom Lewis. As a kid, he'd been forced to resign as an altar boy at the local parish not for dereliction of duty but for a circumstance beyond his control: At sixteen, he was no longer a boy.

A priest prodded Tom toward seminary, a prospect he might have found more enticing were it not for the vow of celibacy. Instead, Tom chose a secular route down the same path: He became a teacher, a

*At the request of the family, the names and some identifying details have been altered.

career choice that didn't sit well with his father. Harold Lewandowski may have Anglicized the family's name upon arriving in America from Poland shortly before World War I, but carpentry, a profession passed along to him by his own father, he considered sacrosanct. From the moment of his son's birth, Harold Lewis assumed Tom would become a carpenter as well.

Thomas Lewis certainly inherited the woodworking genes, assisting his father in the construction of a house before he was a teenager. He was good at it, too. Big, strong and blessed with a sturdy upper body, his fingers and palms quickly acquired the rough-hewn calluses of a man who used his hands for a living.

Rejecting Harold's entreaties, following graduation from high school, Tom entered the service just in time for the end of the Second World War. The army shipped him to occupied Japan to help muster out combat soldiers. When the last of the units departed for the States, he shifted to the transportation pool, an assignment that required him, on occasion, to serve as the driver for General Douglas MacArthur.

Following his discharge, Tom took a job as a draftsman, intending to take advantage of the free tuition offered through the GI bill. Stubbornly, Harold kept pushing for carpentry, a trade that didn't require Tom to waste four years sitting in a classroom. When it became apparent Tom wasn't going to follow the designated career path, his father said he'd settle for the next best thing and encouraged his son, the draftsman, to enroll in engineering school.

Again, Tom balked and reiterated his desire to become an educator. "Teaching is for girls," his father told him. Tom disregarded the ill-conceived maxim and enrolled in the education curriculum at Seton Hall University in South Orange. Then, as was the case when he went on to attain a master's degree from Columbia University, Tom went to school at night. Later, it became a source of pride that he had spent not a single hour in a college classroom by the light of day.

To supplement the GI bill, he carpentered with his father and uncles. On October 3, 1951, the crew had the portable radio tuned to the third and final game of the National League play-off between the New York Giants and Tom's beloved Brooklyn Dodgers.

The part-time carpenter, part-time student was shingling a roof when Giant third baseman Bobby Thomson stepped into the batter's box to face Ralph Branca in the bottom of the ninth with one out, two men on and the Dodgers leading 4–2. When Thomson knocked an 0–1 pitch over the left field wall, Tom almost toppled two stories to the ground.

Three years later, with degree in hand, he began his first teaching job at an annual salary of $2,800. Before he reported for duty, Tom received a pointed reminder from his father that the previous year he'd earned double the amount working as a part-time carpenter.

In the fall of 1953, high school student Constance "Connie" Cuccinello was scheduled to take an early-morning history class taught by a teacher she thought bore a strong resemblance to Abraham Lincoln. During introductory remarks, the teacher announced the class would pursue a vigorous course of study. A heavy dose of papers and quizzes didn't suit Connie, who wanted to be a secretary and had little interest in history.

Conveniently, a scheduling conflict allowed for Connie's transfer to a last-period history class where, she was pleased to discover, the teacher looked nothing like Abe Lincoln and every bit like a young man fresh out of college. Despite a more relaxed learning environment, Connie Cuccinello still had difficulty concentrating on history. The teacher, Mr. Lewis, inquired if the distraction had a particular source. It did: a boy problem. Constance confided her woes to the teacher, a good listener from whose counsel emerged a friendship. She was eighteen; he, twenty-four.

In December, the teacher-student relationship shifted to a new level as they became much more than friends. While the ramifications in 1953 were not what they would be today, the situation was still dubitable. Her father didn't know; her mother, who otherwise would have been her confidante, was dead. His mother knew and pleaded with him to stop; his father said nothing. Somehow they managed to keep it from the rest of the school, a secret helped immensely by the fact that Tom lived, and therefore they dated, in a town a few miles away. The few friends who were aware of the situation demonstrated the depth of their friendship by keeping it to themselves.

The summer after graduation, Connie announced the relationship to her father. He was ecstatic. A teacher! Five months later, on Thanksgiving Day, 1954, they were married.

Children quickly followed: Christopher in 1955, Dean in 1956, Carolyn in 1958. Practicing Catholics, Thomas recognized that he and Connie would need a large house and, as a carpenter, he had the wherewithal to do something about it. In 1959, they moved into the house built by Tom and his father in north-central New Jersey, about fifteen miles west of Manhattan. Two years later, Patricia arrived, followed by Sean in 1963.

Despite the onslaught of urban sprawl in every direction, the quiet street where Tom and Connie put down roots proved a great place to raise children. When the kids, specifically the boys, grew older, the business around the corner, a country club, provided their first source of income.

They never caddied for their father; Tom enjoyed sports only as a spectator. In that regard, following the 1957 baseball season, the O'Malley family broke his heart by relocating the Dodgers to Los Angeles. For five years he pined over lost baseball love, ignoring Connie's exhortations that he switch allegiance to her favorite team, the Yankees. Not until expansion minted the New York Mets in 1962 did he again turn his attention to the national pastime.

Sports were just a cog in a life overflowing with multiple interests and diversions. Before he enrolled at Columbia, the university administered a vocational aptitude test that pointed him toward a career in physics. Connie hadn't the slightest doubt of his ability to excel had he followed up on the test results. It didn't matter if the source came from books, magazines, newspapers, radio or television, every scrap of information he encountered became grist for his spongelike mind.

Musically, he gravitated toward Beethoven and Benny Goodman. Not content just to listen, he had to learn all there was to know about the men and their music. With gardening, it was the same thing. Other gardeners might be content puttering around in the dirt, weeding here and there. Tom picked up gardening books and read every page, ensuring that his tomatoes became a late-summer staple and the blooms on his favorite flower, verbena, a source of bragging rights.

The Bible collection was something that evolved after he picked up a vintage edition abandoned by a nonsentimental family at a secondhand bookstore. Soon, the collection of old Bibles supplemented by fancy new editions expanded to the point that the house looked like a religious bookstore.

Drawing substance from religion and family, it seemed only natural that Thomas Lewis would accumulate Bibles. Quirky and compulsive, he was also enough of a realist to acknowledge that a large and growing family required more than religion for sustenance, and that a teacher's salary could be stretched just so far.

Abandoning the classroom was tough: Tom loved teaching social studies, the daily interaction with the kids, the satisfaction of watching them mature and go out in the world with knowledge learned under his tutelage. But he had a wife and five children to support. Thus began the progression through the public school system hierarchy: assistant principal, principal and, finally, assistant superintendent for business. As an educator, he knew little about business, a shortcoming he rectified by earning a master's degree from Columbia.

Though he moved to an office, his heart remained in the classroom. Once a week he'd return to one of the district's elementary schools to observe from the rear of a classroom. "That's what we're all about," he said in an interview with a local newspaper. "It rejuvenates me."

He also continued as director of student activities, organizing the buses for road trips to football and basketball games, chaperoning the dances and the proms. Even though he was an administrator, yearbooks were dedicated to Tom Lewis and the students still wisecracked with him in the hallway between classes. Next to the family, the high school football team, a powerhouse during the fifties and sixties, became his pride and joy. He served as the field announcer for all the home games and rarely missed a road game.

The urbanized community where Tom worked was, at the time, a city in transition. Heavily industrialized and long a paradigm of ethnic diversity, the city stood in stark contrast to surrounding towns, lily-white bedroom communities that dispatched thousands of workers into Manhattan each morning.

With the opening of the Garden State Parkway, a divided toll road that provided white North Jersey urbanites access to the lower housing prices and larger lawns available in South Jersey, the community's

European ethnicity began to dissipate. As the Italians and Irish joined the southbound exodus to subdivisions along the Jersey Shore, their departures paved the way for an influx of blacks and Hispanics taking advantage of the depleted real estate markets in the hardscrabble cities on the west side of the Hudson River.

While longtime white residents fled the city en masse, Thomas Lewis stayed with its school system, a decision that took its toll as the district struggled to adjust to and accommodate a population of different ethnic constituency. Every night, it seemed, he was out of the house: If it wasn't a school board meeting, it was a city council workshop or a student activity function or a sporting event.

Noticing the stress in several ways, most dramatically in the increased number of cigarettes he consumed each day, his family pushed him to leave the administrative position and return to the classroom. Instead he stuck it out, receiving his just reward when, in 1966, the administration named him coordinator of the district's building and renovation program.

Man and job were perfectly suited. In 1983, he told a local newspaper: "I remember saying to the board of education that I was extremely interested in [the job] as long as we were making it easier for teachers to teach kids. It's important to me that I keep that in focus. Many days this job seems totally unrelated to the education of children."

Project number one was renovation of the high school, followed by the construction of one elementary school and major additions to a pair of others. For his efforts, a local fraternal organization named Thomas Lewis "Man of the Year."

The work was equal parts energizing, rewarding and debilitating. Never the picture of health, Tom had bad arteries that ran in the family: His dad had them as did his uncles. The cigarettes, not to mention the kielbasa he so loved, didn't help the situation.

His first heart attack, at the age of forty, frightened him into quitting smoking and laying off the Polish sausage as he confronted, for the first time, the impact of his mortality on Connie, the children and all that still needed to be learned and understood both spiritually and intellectually. Bad habits die hard, and as the memory of the cardiac arrest faded and the pressures of the job again multiplied, Tom, addicted to nicotine since the age of fifteen, took to smoking a pipe;

inexplicably, he soon began to inhale. The second heart attack, at forty-five, convinced Tom to stop smoking for good. He never did completely swear off the kielbasa.

Work might have compelled him, invigorated him and vexed him, but the center of his universe remained Connie and the kids. During the sixties, while other American nuclear families imploded, the Lewises remained a tight, efficient and loving unit.

While every birthday, every holiday and, indeed, every Sunday was cause to celebrate that bond, Christmas was the pinnacle. In the weeks leading to the holiday, Connie and Tom forbade the children access to the stash of gifts by temporarily barricading the upstairs hall leading to their bedroom. As Christmas drew closer, they could barely navigate the bedroom, what with all the unwrapped gifts lying about. Unlike some practicing Catholics, who exchange gifts after midnight Mass, they celebrated on Christmas morning, which began with the children creeping down the stairs to see what Santa Claus had left under the tree.

Afterward, the real festivities began: At the Lewis home, Christmas and Easter were transcendent; more than mere holidays, they became rituals steeped in food, family and religious tradition. Connie Cuccinello Lewis handled the Italian side: poppy seed cake, biscotti, crustata; her husband provided the Polish delicacies: babka, pierogi and, of course, kielbasa. When it came to Christmas breakfast, Polish tradition won out over Italian with Tom serving up kielbasa, homemade horseradish and kilebaka bread.

The Lewises' oldest son, Chris, often felt sorry for his friends whose own families, compared with what transpired at the Lewises', seemed to celebrate Christmas hardly at all.

In the Lewis household, Sundays were but a mini-version of Christmas and Easter with mandatory attendance at dinner, not that dispensation to skip a family meal was ever an issue. The grip of the familial knot was best exemplified by the crisis that erupted after Chris received formal notification that he'd been accepted into the premed program at a prestigious college in Pennsylvania. The news that their firstborn would be attending a Catholic university thrilled the Lewises, an enthusiasm tempered by the geographical distance about to be placed between Chris and the rest of the family. Located two hours from their home in New Jersey, to the Lewises, the school might just as well have been at the other end of the earth.

If the family lay at the center of Thomas Lewis's universe, the Roman Catholic Church was his solar system. His was not a passive belief, limited to perfunctory attendance at weekly Mass: Long before there were schools to be built and flowers to be grown and history books to be devoured, there had been the Church; it was as much a part of him as his size 11 feet and the calluses on his hands. The bond between Tom and his religion, instilled at birth, had never diminished, and in fact, with each passing year it grew ever stronger.

When it came to faith, Tom never proselytized, nor did he hold his convictions in abeyance, providing a full account of his belief to anyone demonstrating even the slightest curiosity. He didn't claim to have all the answers, but he knew what had worked for him, Connie and the family. And he wasn't the least bit reticent about letting others in on the secret.

To that end, in the early 1970s, Tom persuaded Connie to join a Marriage Encounter group; it didn't take long before the Lewises became the organization's facilitators. Fighting through the reluctance to discuss marriage in the context of spirituality, Tom guided the couples through the process. Guilelessly using his own marriage as an example, he gently prodded the tentative to become forthright, helping the supplicants to find, through faith, mutual solutions to individual problems.

"Tom and Connie had a way of centering on God in their life. It was vitally important to them," said a priest and family therapist who helped the couple to counsel others.

Drawing on unabashed faith, values, intellect and an innate sense of compassion, Thomas became the beacon to whom the rest of the group turned. He was so confident and so sure of himself that, if she hadn't known better, Connie would have sworn marital counseling was his life's work.

Overshadowed by Tom, Connie felt as though she were only along for the ride until she came to the realization that while her husband connected to the others from a position of intelligence, hers was a bond founded on speaking from the heart. Thomas concurred, telling her, "I am the vine, you are the leaves." Connie knew he was right: Tom is the vision, she often thought, and I am the dream.

As word of the husband-and-wife team spread through the Catholic hierarchy, the Newark Diocese called and soon the Lewises were co-chairing the Marriage Encounter movement for the entire state, organizing workshops and weekend retreats, imparting the lessons they'd learned in their own marriage.

One thing Tom never learned for his own benefit was how to pace himself through the enumeration of family, faith, vocation and avocation. Not coincidentally, the inability to slow down took its toll, specifically in the matter of his own heart. Weakened by arrhythmia, he realized something needed to be done lest he suffer the same fate as his mother and uncles on his father's side, all killed by heart disease while in their fifties.

The doctors in New Jersey dispatched him to specialists at Columbia-Presbyterian Hospital in New York who informed him his only recourse was a procedure known as a coronary bypass. Tom wavered: Bypasses in 1981 were not the common medical procedure they are now.

When the price for reticence was a medical prognosis that grew increasingly grave, he gave the go-ahead. Following the decision, he summoned his son, Dean, to his bedside. If I don't make it, Tom whispered to his son, I want you to become a priest.

Before the surgical team could proceed, they first needed to stabilize the patient to prevent a third heart attack—this one, undoubtedly, fatal. Through it all, Tom remained upbeat, teasing the hospital staff and volunteers, bolstering the family with inside jokes. To Connie, Tom's wait to gain the strength necessary to withstand the surgery seemed interminable. Finally, on President's Day, five veins and arteries were harvested from his leg and grafted into his chest.

Secondary pneumonia—a common side effect of open-heart surgery—set in while Tom was still in the recovery room, causing fluid from the lung to spurt all over the bed when an aspirator malfunctioned. What next? Connie asked herself and pitched in to help a befuddled intern clean up.

Sparing Dean the onus of fulfilling a deathbed promise (or the guilt of not granting his father's wish), Thomas Lewis made it.

When the slow recovery began, Tom's first concern was for the family's suffering on his behalf. "I should have died after the first heart attack; I could have left you a rich widow," he told his wife.

Given the state of his health after the surgery, death became more than a joke. The prospect that she might lose her husband sooner rather than later began to weigh on Connie Lewis. She had little doubt he would go first; to think about when it would actually happen was too painful.

Still, at Tom's behest, they began to plan, with Tom convincing Connie to join him in donating their bodies to science; medicine, he told her, stood to gain a wealth of knowledge from his body. Connie's mother had died when she was very young; she recalled little of the actual death, but had never been able to purge the aftermath from her memory. The image of her mother resting in an open casket remained a particularly vivid and grotesque memory. Contrary to the suggestion of well-meaning friends and relatives, Connie's mother did not appear as though she were sleeping: She appeared to be dead. The lifelong residual impact was such that, even in adulthood, Connie Lewis avoided wakes.

Thus, Connie and Tom's arrangement to bequeath their bodies was predicated on more than an emotional reaction to his spate of medical maladies. More than most, Tom and Connie knew exactly what they were getting into: In a close family, everything is shared, and it hadn't been that long since Chris, following up on a lifetime goal to become a physician, had been a first-year medical student keeping them abreast of his progress through the mandatory gross anatomy course.

None of the children were particularly surprised when their parents informed the family of the decision; it was, after all, a natural extension of their lives. On May 21, 1982, they signed the papers; two days later the envelope containing the forms was delivered to the cluttered sixth-floor reception area outside the office occupied by the chairman of the Department of Anatomy at the New Jersey Medical School in Newark.

The department secretary, Essie Feldman, opened the envelope, reviewed the information and made sure the forms were properly signed. It was a typical day for Feldman, meaning there were class schedules to coordinate, telephone inquiries from potential donors that needed to be addressed and letters to be typed for the chairman. Feldman placed Thomas and Connie Lewis's paperwork in an alphabetized file cabinet along with forms submitted by hundreds of other donors and then moved along to her next assignment.

Life went on, too, for the man with the name now in Feldman's file cabinet. With the school system requiring Tom's continuing attention, there was no thought of retiring; his kids were going off to college and he had to pay the tuition, despite the multiple surgeries that depleted his energy and left him vulnerable to other problems, including skin cancer and bleeding ulcers.

The next major medical crisis arose from poor circulation: the blood vessels leading to his left kidney were occluded, causing his blood pressure to skyrocket. As the arterial sclerosis slowly weakened the kidney, the Lewises embarked again on the odyssey of medical consultations, which resulted in the determination that the left kidney had to come out.

After the surgery was performed, in 1985, Tom told Chris that a group of future medical students would one day receive the lesson of their lives when they encountered the myriad medical anomalies inside his body. "They're going to be lucky to get me," he joked.

This time, at least, he didn't need to rush back to work. In June of the previous year, Tom had retired from the school system, telling a reporter: "The first reason is because I've had some health problems that are a concern. I'm taking some advice that I may not have accepted a few years ago."

The bittersweet leave-taking provided a forum, allowing him to postulate publicly about the future of education. "Most of us who were trained as teachers in the late forties and early fifties entered a profession where the focus was truly on education of the youth. The schools we entered were primarily educational establishments. Today, we are psychologists, social workers, parents. We are expected to do a lot more than we were a generation ago," he said in the interview.

"Parents don't have time for children, whether by choice or not. The pressure of the economy forces both parents out to work . . . And don't overlook the feminist movement. It encourages mothers to be more than mothers. . . .

"I still believe that the greatest job a woman can aspire to is motherhood, once the decision is made to become one. I have no problem with women who don't want to be a mother, but the abdication is what's giving us most of our problems today."

Though disillusioned, Thomas Lewis let on that his years in urban education had not left him intractably discouraged. He told the reporter: "I'm not looking at retirement as an end, but as a beginning," he said with a playful smile. "Maybe I'll become a teacher again. That's what I started out to do."

Selling the house in north-central Jersey, Tom and Connie moved south to a retirement home an hour north of Atlantic City. She started selling real estate; he returned to the classroom, part-time, as a substitute.

After three decades working on behalf of other people, retirement suited Thomas Lewis. Voraciously compensating for lost reading time, he took in each day's mail, which brought magazines on a variety of topics: *Time, Newsweek* and *U.S. News & World Report;* history, art, music, religion. When he wasn't reading a magazine, Tom had his nose buried in a book, mostly history, but also popular nonfiction. Not content just to read, he was kept company by the television. With a single exception, *Jeopardy!,* it was tuned, always, to CNN. Addicted to *Jeopardy!,* Tom knew all the answers, including those the contestants couldn't answer. Whenever she spotted a notice for local auditions, Connie urged her husband to try out for the show. A smile was Tom's standard response.

Here today, forgotten tomorrow is the lot of substitute teachers. Not Mr. Lewis, who overcame his status as a substitute by bringing to each classroom an enthusiasm equal to that which had so impressed a certain girl seated in the front row of the first history class he ever taught, thirty years before. It amazed Connie Lewis that every time they'd run into kids from the high school at the convenience store or ice cream shop, they'd not only engage her husband, they'd literally embrace him.

These were good years, highlighted by the addition of grandchildren, with whom Poppy enjoyed crabbing on nearby Barnegat Bay, seamlessly integrating them into the family traditions. With grandchildren, Christmas and Easter took on new meaning and, if anything, the holidays became even more special.

The Church also benefited from Thomas's retirement as he and Connie, true to form, proved themselves no ordinary congregants. More than happy to tap into the Lewises' expertise in the field of marriage and faith, the parish priest persuaded Tom to head up its program to introduce converts to the history and rituals of the Church.

Before long, there were again too few hours in each day to accommodate the diversity of Tom's passions. When he wasn't at the church he was at school, and when he wasn't at school then chances were good that he was volunteering at a hospital in a neighboring county.

From a purely personal standpoint, Thomas had a dual interest in the medical center's mission: his cardiac surgery and the work being done by Chris, who'd become a cardiologist.

As with everyone who came in contact with him, the hospital recognized Tom as a man who could get things done and quickly put him in charge of its Zipper Club, an organization dedicated to assisting patients in the aftermath of open-heart surgery. To further the cause, the hospital recruited the Lewises to reenact the rituals of cardiac surgery for an informational video that the club distributed to prospective patients. The reenactment, featuring role-playing by real doctors and nurses and, in Thomas's case, a real patient, gave Connie the creeps. For Connie, the scenario was all too real, especially as it became clear the role was about to be reenacted in real life.

Epitomizing the nascent procedures performed in the 1970s and early 1980s, as the survival rate of bypass patients like Tom increased, medical science began taking note of relapses caused by blood vessels transplanted from the leg that were weakened and narrowed due to a number of factors, not the least being patients who failed, as had Tom Lewis, to abide by a strict regimen of healthy diet and exercise.

In 1991, the deterioration of Tom's bypass had become serious enough to warrant his return to Columbia-Presbyterian for a second bypass, this time using veins and arteries harvested from the thoracic cavity. While it took only a few weeks for Tom to resume his familiar routines, the second operation left Connie Lewis feeling increasingly as though she was living on the edge. Over the next several years, Connie seized up at the least indication that her husband's stamina was waning. This is it, she'd think, the moment I've been dreading.

Still, when Tom fell ill during Easter Week 1997, seven days after returning from their annual Florida vacation, Connie was not overly concerned. He had stomach flu, a virus that had nothing to do with the heart.

Feeling no better by Good Friday, Tom returned to the doctor to report that the regimen regarding food, liquids and rest prescribed on Monday had resulted in little improvement: The previous evening, at the Last Supper, he'd eaten practically nothing. Give the intestines a break, the doctor told him, limiting his diet to Gatorade and Mylanta.

It seemed to work. Feeling stronger Saturday, Tom spoke of attending Mass on Easter. Then, about 6:30 Saturday night, came the onset of chills. When he began shivering heavily, Connie pulled out the *Mayo Clinic Family Health Book*. Consulting the chapter on Influenza, she ascertained her husband had every symptom: diarrhea, chills, nausea, fever. Chris, speaking to her by telephone, concurred with the diagnosis and directed Connie to cover Tom with blankets, advice she'd already obtained from the health primer.

With Chris still on the phone, Connie checked her husband again; the chills were subsiding and Tom was sound asleep. "The worst is over; when we wake up Easter morning everything will be fine," Connie told her son.

Connie Lewis awoke Easter morning at 6:45 and peeked over at her still-sleeping husband. Connie put her hand on his forehead. Cold. Deciding to let him be, Connie climbed out of bed. Halfway to the bathroom, she halted abruptly; her eyes swung back to the bed. Cold. She looked at Thomas lying under the covers and then toward the floor and saw something out of place: Thomas's foot dangling lifelessly off the side of the bed. The moment so long dreaded had come.

Connie first contacted Chris, who instructed her to call 911 as he flew out the door for a mad, half-hour dash down the Garden State Parkway to his parents' home. Next, she called her daughter, who lived nearby. The ambulance showed up, and the first-aid volunteers rushed in and began CPR as Connie stood in the corner of the room and watched. There was really no point, of that Connie Lewis was absolutely certain.

The ambulance took Tom to the nearest hospital, to which Connie summoned the rest of the children. They arrived just before noon on Easter Sunday. The family was ushered to a small room where their husband and father and grandfather lay. They hugged him and hugged one another and they wept. This went on for nearly an hour.

Finally, a nurse entered. It's time, she told them as gently as possible. And so they said good-bye.

At the memorial service held six days later a bagpiper played "Amazing Grace" in honor of a Pole with an affinity for Irish bagpipes. All the children and most of the sixteen grandchildren participated in the service. God and family were at the center of the hour-long memorial; Tom would have loved it. The church was packed, reminding several in attendance of the High Mass accorded top Church officials.

Gone in body, Tom remained in spirit to the extent that, at the party following a grandson's middle school graduation a few weeks later, the family continued to speak of him in the present tense.

One day, as spring turned to summer, Connie and one of her daughters-in-law, Susan, decided to pay a visit to the hospital where Tom had volunteered. As they passed a bakery on the way home, something clicked: It was the place where Connie's husband used to stop to buy PowerBars and pastries on his way home from Zipper Club meetings.

Connie made a U-turn and returned to the bakery, having no idea what she would say as she walked through the door. When the cashier came to the counter, Connie introduced herself, explaining that her husband, Thomas, had been a customer.

Thomas Lewis, said Connie, perhaps you remember him? He passed on, I thought you might like to know. Stunned, the woman excused herself to the rear of the store, reemerging a moment later with the proprietor, both of them copiously weeping.

No, they hadn't heard, but they'd feared as much when Thomas, they'd never known his last name, had failed to make his customary visits to the bakery over the past few months. Sobbing, the two women embraced Connie and Susan, who stood there, dumbfounded: There they were, in the middle of a county where they knew not a soul, consoling two people they'd never laid eyes on who were anguishing over a man Connie and Susan loved dearly, a man whom the two strangers knew but cursorily.

Even in death, Connie found Tom all over the house, in the Bibles stashed in every corner, the family genealogy grid left on the hard drive of the computer, the handwriting in his phone book.

Eventually, the death certificate arrived, bringing with it a measure of finality. The official record filed with the County of Ocean in the State of New Jersey stated that Thomas R. Lewis died early on the morning of March 30, 1997. He was sixty-nine years old. Cause of death: acute myocardial infarction. It said nothing about him being a husband, father, grandfather, devout believer, teacher and a carpenter extraordinaire.

While time didn't stop for Tom's family, Christmas that year, as it would be for all the years to follow, was not the same. Nor were the subsequent birthday parties, holidays and family Sunday dinners, at which, all agreed, a key ingredient was missing.

In the two years after Tom died, at the times when she missed her father the most, Carolyn Lewis O'Donnell would often take account of the lasting impact his life had had on her and everyone he'd encountered. And each time she did so, Carolyn reminded herself that, in a strange way, her father was still hanging around, albeit up in Newark. That, Carolyn thought, was kind of cool.

CHAPTER 4

———————————•———————————

While two weeks passed before many in the class summoned the courage to gaze upon the faces of the men and women they were about to dissect, no such recalcitrance evidenced itself at Table 26, where Sherry revealed the cadaver's gender the moment the towels were removed. Jen's gaze moved from the hairline of the man to the twine stitching a three-inch incision on the left side of his neck.

"Embalming," Udele explained.

"They all have it," said Sherry, shrugging toward Table 25, another table that had removed the towel from the face.

"Take a look, see what we're dealing with here," said Jen, turning to Ivan.

"A man of few words," Sherry observed, wishing it weren't so. Sherry much wanted to know about his wife, children and grandchildren, an innate curiosity extending also to what he did for a living, for pleasure, his beliefs and values. Above all, Sherry shared with her classmates the overriding question: What force compelled him to sanction what was about to be done to his body?

Accustomed to ministering to the needs of the uncommunicative— the men and women arriving at the various ICUs and trauma units where she'd served were generally unconscious—Sherry knew the eventual visitation of family and friends would provide the nursing staff some insight about a patient.

This differed in that, beyond his anatomy, the man at Table 26 would forever be a stranger. Sherry would never know him as he had walked the earth in the person of son, husband and father; nonetheless, she would soon know him more intimately, in a very different sense, than all who had once loved him.

Tall in stature, the man had a prominent nose, large ears and a receding hairline. He reminded Ivan of his father. The eyes were shut tight as was the mouth; for that the students at Table 26 were grateful. Pursed lips were preferable to a mouth contorted in what med students call a "death scream."

Breaking through the veil of secrecy, there was one aspect of Number 3426's life the anatomy department did make available: how he died. The causes of death for the eleven cadavers in Lab C were posted on a stanchion at the front of the room, directly above a notice assuring the students that each body had tested negative for hepatitis B and HIV.

"COPD," Jen reported after a quick trip to the stanchion.

Sherry began looking for the outward signs of chronic obstructive pulmonary disease and immediately found evidence: a barrel chest, blue-gray fingernails.

"A smoker," she said, dismissively. A health nut, Sherry disdained smoking. Still, the prospect of working with a nicotine addict enticed; a life of smoking promised some very interesting postmortem aberrations.

Searching for further superficial documentation, Jen and Udele began to methodically peek under the five towels covering the rest of the cadaver, reflexively averting their gaze when they reached the towel covering the genitalia. Removal of the last towel, the one spread across the feet, revealed an orange plastic bracelet secured to the left ankle bearing the number 3426.

Once white and billowy, a decade of exposure to phenol and decomposing bodies had left the towels a threadbare orange-yellow. So many had been destroyed during the last industrial-strength washing that Roger Faison asked the school for replacements; the school ordered the mortician to make do another year. "What can I say? It's a state institution," Faison said with a shrug.

The towel predicament became so dire that, while covering the cadavers in Lab D the week before Christmas, Faison ran out. When enough new towels miraculously materialized to rectify the situation, Faison rejected them; he'd been around academia long enough to realize that one lab with new towels and three labs with old towels were the stuff of conspiracy theories. Reluctantly, Faison opted for uniformity, arranging for intact towels to cover the larger cadavers and using the tattered remains to shroud the more diminutive.

Udele replaced the last towel. "He looks like he's in good shape," she proclaimed.

"I just want to look, I'm so curious," said Jen, a kid with a new toy, studying Number 3426's face and chest, stopping at the abdomen. "Do they vacate the intestines?" she wanted to know.

"Uh, I don't think so," Udele replied.

Unexpectedly, the lights dimmed and the television screen above Table 26 flickered to life. A disembodied voice with a vaguely European accent boomed Oz-like from speakers mounted in the ceiling. "Good afternoon," said the voice.

"This is eerie," said Jen. In the not distant past, instructors conducted prelab lectures in the traditional manner, scrawling illegibly across the labs' whiteboards with black felt markers. Then, in 1998, modern technology arrived in the form of a specially outfitted PowerMac graphics system that allowed a simultaneous broadcast of the prelab lecture to television sets mounted in all four labs.

Using an electronic pencil to diagram a photograph of a cadaver's chest, the instructor responsible for the prelab lecture outlined the details of the first day's dissection, beaming it, via the computer age's version of an overhead projector, to the 181 first-year students assembled around television sets placed throughout the lab.

Initially, the introduction of multidimensional computer imagery led many medical educators to believe high technology would render traditional gross anatomy laboratories obsolete. Instead, anatomy turned out to be immune to technological advance. The imagery programs are still in use, but only to supplement the activities in the lab. In understanding the anatomical form, medical schools concluded that eliminating the odor associated with dissecting human bodies was about the only advantage to mouse-click edification. In anatomy, to learn is to cut; there is no substitute for the real thing.

"Identify the clavicle bone and sternum," instructed the voice, that of Dr. Zolton Spolarics. As Table 26 would soon discover, Spolarics, a native Hungarian, was the primary instructor assigned to Lab C for the first of the three units comprising the gross anatomy laboratory curriculum—the thoracic cavity; back and spinal cord; shoulder, arm and hands. Normally, the instructors, in teams of two, rotated between the four labs for each of the three units. Because of scheduling conflicts and illness, however, Spolarics basically had to go it alone in Lab C during Unit I, assisted occasionally by other instructors brought in to help out.

Spolarics continued, "Remember to find the landmarks. This is very important. Today, please do not disturb the deeper structures. This, also, is very important."

In Lab C, only Sherry took notes. Medical school being nothing if not autodidactic, everyone should already have been familiar with the thrust of Spolarics's lecture. Prior to the first day and every subsequent day of lab, the anatomy department assumed that each member of the class had adequately prepared for the next stage of dissection. If they hadn't, it wasn't the department's problem.

Preparation began with course coordinator Dr. Nagaswami Vasan's ninety-three-page course syllabus. Orange-bound and meticulously detailed, the syllabus laid out, day by day, the fourteen-week journey through the human body, including course objectives, lecture topics, lecturers, lab assignments and exam schedules.

The syllabus, Vasan emphasized in the introduction, was only a guidepost. For erudition, the students turned to several other sources. *Clinical Anatomy for Medical Students* by Richard S. Snell correlated function and pathology to the structures revealed during the dissection. And while several atlases received the official blessing of the anatomy department, the favorite of the students was Frank H. Netter's superbly illustrated *Atlas of Human Anatomy*.

When it came to a textbook that could precisely detail the intricacies inherent to the art of dissection, the department offered no alternatives. In the gross anatomy lab at NJMS and, indeed, at medical schools around the world, *Grant's Dissector* is the bible. It is a bible that traces its origins back to Andreas Vesalius, the sixteenth-century Belgian-born physician credited with moving medical science into the modern age. In immortalizing the blueprint of the human body, Vesalius had a collaborator, Jan Van Calcar, an illustrator who complemented the science of medicine with details of the human anatomy that have so withstood the test of time that they are said to be among the prototypes for *Grant's Dissector.*

Twelve minutes into the lecture, Spolarics set aside the electronic pencil. "Do not cut below the rib cage," he warned solemnly. "You will eventually cut open the abdomen. But you will not do this today. It is smelly. It will dry out and make things very bad for you later. Good luck."

The lights came up, and an expectant banter filled the room about the collective act about to be performed. Those most uncomfortable avoided entirely the topic of dissection, adhering to a grand medical school tradition grounded in the human condition: fear of the unknown.

"They really don't know what lies ahead. They don't know what they're going to find," said Dr. Sandra L. Bertman, a professor of medical humanities at the University of Massachusetts, and an authority on the relationship between medical students and death.

Or maybe they do. From the moment a future physician sets his or her sights on a career in medicine, this step—dissecting a human body—looms between them and the objective. Passage through Medical Gross and Developmental Anatomy, one medical student said plaintively, is "paying your dues for medicine. It is the bridge you have to cross" in order to become a physician.

Mesmerizing, appalling, emotionally depleting and all-encompassing in its scholarship, it is a rite unique to schools of mortuary science, dentistry, physical therapy and, foremost, medicine.

As NJMS student Leah Schreiber put it, "The only people who do this are medical students. And psychopaths."

Romantics at heart, the public likes to believe that in every physician the seed for a career in medicine is planted early and germinated through the healing of a broken childhood limb by a compassionate general practitioner, the inspiration of the physician who interceded between death and the life of a loved one or, conversely, by the witness paid to a noble failure in the attempt to prevent the loss of a father, mother, grandparent, sibling.

At Table 26, only one student, Ivan Gonzalez, approximated that ideal. Udele Tagoe, born in 1977 across the street from where, twenty years later, she would attend medical school, could not remember a time when she did not want to become a doctor. Neither could Sherry Ikalowych. Persuaded not to apply to medical school as an undergraduate, Sherry got sidetracked by career, marriage and motherhood before proceeding with the original plan at the age of thirty-nine. On the other hand, Jennifer Hannum's medical aspirations came four years into a five-year pharmacy program at Rutgers University. As with Sherry and Udele, there had not been a defining event or experience that propelled Jen toward medical school.

Then there was Ivan. Plagued by asthma, he had spent his childhood ricocheting from one health clinic to another in his native Managua searching for relief from the congestion and wheezing that left him desperate for oxygen. Inhalers didn't help, nor did pills or injections. Much of the problem lay within: Ivan used the inhalers as pharmaceutical squirt guns to terrorize friends, brothers and sisters, and as often as not, he obscured the benefits of the antibiotics by refusing to adhere to the attendant dietary restrictions.

Ivan could recall neither the name of the clinic that provided the elixir nor the physician who prescribed it. What remained ingrained was the exhilaration that came from knowing, henceforth, that his breathing needn't be encumbered by a debilitating respiratory ailment. From that point forward, Ivan wanted nothing more than to heal as he had been healed and nothing—not revolution, poverty, language barriers or doubt, his own and others—could intercede to prevent the dream from being realized.

Five days into the first month of the last year of the twentieth century, the dream landed Ivan in a figurative place through which every man and woman who has sworn to uphold the Hippocratic oath has passed. A place where life is seized from death. A place that, ultimately, would provide Ivan and his classmates what one former NJMS student called "the very foundation of our medical knowledge."

The syllabus directed Ivan, Sherry, Udele and Jen to page seven of *Grant's Dissector.* The dissection of Number 3426 would begin with the thoracic cavity. Not every anatomy department starts with the chest. Some schools, the State University of New York at Stony Brook, for example, start students on the back and spine in an effort to minimize the initial trauma by hiding the face. As the lab partners at a SUNY-Stony Brook dissection table learned a few years ago, the best intentions of the school often result in a problem of a different nature.

Based solely on the length of the cadaver's hair, the four partners at one table concluded on the first day of lab that the body was that of a man. They named him Frank, "because his job was to uncover to us all the 'true' intricacies of the human body," one of the lab partners wrote of the play on words. (At the conclusion of its anatomy course, SUNY-Stony Brook requires its students to record their impressions in a fictional or nonfictional essay.)

After the table had dissected Frank's back, they prepared to examine the cadaver's thoracic cavity. The student wrote: "Naturally, it required us to turn Frank over into the supine position. This would be the first time we would see his huge fat belly in all its splendor. None of us expected the surprise that awaited us. Once we turned the cadaver's obese body over, we were shocked to learn that Frank was not what we thought he was. Frank was a woman!" That afternoon Frank was rechristened Fran.

While no similar surprises awaited Table 26, before they could begin dissecting they encountered another delay. This time Sherry was the cause.

When Vasan asked Sherry to share her expertise with a predissection demonstration on central intravenous line replacement in the neck, she readily accepted the invitation. In addition to being a nurse, she was also a teacher. A decade before, after receiving her master's, Sherry had remained at Columbia University, imparting there the lessons learned at night in the trauma unit at UMDNJ-University Hospital, the medical facility attached to NJMS. Sherry so flourished in the give-and-take of the classroom that, by the time she taught her last anesthesiology class at Columbia, she had risen to the rank of assistant professor of nursing. Now, years later, Sherry took note of the prevailing body language of her medical school classmates and realized no one had the slightest interest in the intricacies of inserting a central intravenous line.

Ignoring the whispered exchange behind her ("Why is she doing this?" "Because she's a surgical nurse"), Sherry broke open the medical kit and proceeded expeditiously. "You need to move the head to the side a little," she began and then laughed as the attempt to move Number 3426's head to the side ended in utter failure. Sherry shrugged, allowing she'd never tried the procedure after the onset of rigor mortis. "Well, it works with someone whose head is, um, a little more viable," she smiled.

Two minutes later, the demonstration over, Sherry placed the tubing and scalpel back in the IV kit and headed for the other labs, where the response was equally indifferent. Still, she had a payoff: As an honorarium, Vasan slipped her two copies of *Grant's Dissector*. One she took home, the other, to the delight of her cash-strapped lab partners, she donated to the table, thus sparing them the ignominy of being in the debt of fellow student Reuven Bromberg.

Prior to lab, Reuven circulated from room to room announcing that he was making a run to the bookstore and offering to purchase extra copies of *Grant's*. "I'll put it on my credit card. You can pay me back later," he proclaimed.

"He must have some limit on that card," Jen marveled.

Sherry thought there must be a catch, surmising that Reuven, a few dollars short, had hatched a scheme to, in effect, borrow against a credit card using his classmates' cash as collateral.

Until Reuven set himself up as the financial middleman between his classmates and the bookstore, many students hadn't yet realized that, in addition to a personal copy of *Grant's*, they would need to share the cost of another $35 edition with their lab partners. The fifth would be a lab copy. And in the lab the book would remain, absorbing the odor and whatever else happened to get splattered upon it. At semester's end, the communal books overflow the lab trash cans, receptacles placed under signs with the stern warning DO NOT PUT HUMAN TISSUE IN GARBAGE PAILS.

Sherry slipped away to complete her tour and an uneasy calm settled over Table 26. Udele picked up a scalpel and inserted one of the blades that Sherry, who'd worked up to the last minute completing her final hospital shift at eight o'clock that morning, had received as a farewell gift from one of the trauma surgeons. Setting the scalpel on the towel covering Number 3426's abdomen, Udele placed the first two fingers of her right hand over the cadaver's sternum and gently probed the bones under the neck.

The key to dissection, whether animal or human, is to first locate the anatomical landmarks. *Grant's* recommends that students begin on the surface anatomy with the thoracic vertebra and then locate the sternum, the articulated vertebral column and, finally, the jugular notch. Reading aloud the pertinent points from *Grant's*, Jen guided Udele through the process. The hands of Ivan, Jen and Udele were covered with two pairs of latex gloves. Between the two layers, they'd squirted pink dispenser soap, a supposed preventative against the first-year smell permeating their hands.

As Udele searched for the landmarks, Jen touched her own clavicle. "It's amazing how much I take for granted that I don't know it all," she said. Using her sternum as a guide she took her other finger and located Number 3426's jugular notch. "It should be right here," Jen said.

"I think maybe he has some fat, but we won't know until we open him up," said Ivan, until then a silent observer.

The moment at last had arrived.

Be it face or arm, leg or neck, the procedure for beginning a dissection, any dissection, is quite specific. For their first procedure, Jen turned to page ten of *Grant's* and read aloud: "With the cadaver in the supine position (face up), commence with the skin incisions. Do not cut too deep. You may inadvertently damage superficially positioned nerves—" Udele's right hand rested on Number 3426's leg. Her eyes on the scalpel and her face a mask of raw concentration, she appeared not to hear a word Jen was saying.

"Can I give you a demonstration on how to reflect the skin?" A clipped British accent interrupted Jen in mid-sentence.

Cambridge-educated instructor Dr. Michael Rose, rotated into Lab C to assist Spolarics during the first week, stepped up to the table. Deferentially, Udele moved aside, her composure crinkled by the realization that Rose's query had been purely rhetorical. The honor of making the first cut at Table 26 would not be hers. "Everybody has favorite dissection tools," Rose continued, oblivious to Udele's disappointment. "Mine are the scalpel and sawtooth forceps."

Picking up the scalpel from where Udele had placed it, in a single stroke Rose sliced halfway down the sternum from the nape of Number 3426's neck. In a second singular movement, Rose then cut a vertical incision halfway across the cadaver's left breast. Jen, her eyes agape, looked at Ivan, then Udele.

Rose hunched over the cadaver and used the scalpel to pull back the skin. "The reflection [of the skin], you see, is not unlike peeling an orange, is it?" he said, gently scraping away the top layer of fatty tissue below the superficial fascia.

"Keep it shallow, not too deep. We're not sure what is underneath there. With each stroke of the blade you go deeper and deeper," he advised. The instructor pulled himself to full stature to admire his handiwork. "There you have it," he said proudly just as Sherry returned from her teaching stint.

"Oh, I missed the first part," she moaned.

In his haste to get Table 26 up and running, Rose failed to notice the table was missing a lab partner. Now, with all present and accounted for, some parting remarks seemed fitting. He plucked a strand of excess skin from the table and told them:

"Remember, the fascia goes in the bag and keep the flaps [sheets of attached skin] covered with the towels when you're finished or they'll start to get smelly. Throughout the dissection, you'll be going from the known to the unknown. Use the bones as landmarks." Rose paused to again admire the incision.

"And away we go," he said, dramatically placing the scalpel back on Number 3426's towel-covered abdomen.

In a room filled with forty-four students and two instructors, the three women and one man at Table 26 were oddly alone, pondering what needed to be done. Failure to consummate the next step meant all they'd done to reach that point—sweating premed courses, the Medical College Admissions Test (MCATS) and the agonizing wait before learning they'd been accepted by a med school—would be for naught. The next fourteen weeks would determine if the dream that three of the four had harbored since childhood would live or die.

For Udele, Ivan and Jen, the dream was not theirs alone: They were the first in their families to attend college. The personal biographies the lab partners brought to Table 26 stopped just short of extraordinary, thematically embracing escape from political revolution and war, escape from the unforgiving streets of the very neighborhood where they now stood and escape from the bounds of sexual stereo-typing.

Yet, somehow, they'd all arrived at this place and point in time, four students with nothing in common, about to share in an experience which, even should they never see one another again following graduation, would bond them forever. As Rose drifted away, they stood alone and they stood as one, with nothing left to do but cut.

Jen studied Rose's incisions and used her hand to probe for the second rib, the bony landmark to which the incision needed to be extended.

"It's really hard to find anything, he's so stiff," said Sherry. From her pocket, she produced a purple felt-tip pen and pushed it in Jen's direction.

"We can use this?" To Jen, the prospect of drawing a pattern on a body, in purple ink, no less, seemed disrespectful. Dignity for the cadavers had been a mantra invoked solemnly and repeatedly by Dr. John H. Siegel during the previous day's introductory lecture.

"Why not?" Sherry replied glibly. "Surgeons do it."

Using the pen, Jen carefully marked the path. She picked up the scalpel and drew a deep breath. "Here goes," she exhaled.

"Just do it real superficially at first," said Udele, scalpel at the ready, about to duplicate Jen on the right side of Number 3426's chest.

Certain she was cutting too deep, Jen haltingly prolongated Rose's incision, looking up periodically for encouragement from Sherry, who happily obliged her.

No such hesitation evidenced itself on the other side of the body. Udele's prelab misgivings that the lessons of the previous summer had been lost in the academic haze of the first semester evaporated as, for the first time in six months, she picked up a scalpel knowing exactly what had to be done and, with single-minded confidence, how to do it. Mindful that her lab partners were looking to her as somewhat of an expert made it all the more important that she get it right.

Udele well understood how unsettling the steely calm with which she approached the task might appear to the world outside the medical community. Aware of the reaction when casual acquaintances learn that spending fourteen weeks in a room filled with dead bodies is one of the requisites of becoming a physician, most medical students tend not to discuss what transpires in an anatomy lab with anyone other than close friends and relatives.

Even those at ease with the concept of keeping company with cadavers might have trouble with the next step required of future physicians. For cutting through the skin of a fellow human being, alive or deceased, is not what most would consider a natural act.

On the afternoon his table prepared to make their first cut in 1997, NJMS graduate Dr. Alan Nasar was consumed with anxiety, unable to shake the memory of a high school science project that had required him to prick his own finger for a blood-sugar test. Around him, fellow students performed the procedure effortlessly and with minimum anguish. Nasar could not do it, halting each stab of the lance a millimeter short. Sheepishly, he persuaded a classmate to draw the blood drop.

Years later, as human dissection was about to move from supposition to reality, the image weighed so heavy on Nasar that he eagerly—too eagerly, he thought—deferred the first cut to a lab partner. When, finally, the scalpel wound up in his hands, Nasar summoned every ounce of resolve and, to his astonishment, discovered he didn't just tolerate the sensation of a scalpel slicing through the skin, he actually enjoyed it.

Nasar was certainly not alone in his enlightenment. As the members of the Class of 2002 also discovered, dissection—the act of cutting—can be intangibly addictive.

Sherry understood the phenomenon only from a secondhand perspective. In a lifetime spent on the outside looking in, or more specifically, the topside looking down, the nurse-anesthetist had long envisioned the day when she would be the one wielding the scalpel. Sometimes it seemed as if it would never come. Now it had. Over two decades after her mother and other forces persuaded Sherry that she and medical school were incompatible, she took the scalpel from Jen. Before lowering the blade she lingered, savoring the moment.

Then, deftly and with an experienced stroke gleaned from years of observing surgeons do the same, she started to scrape at the deep fascia with unbridled intensity. Normally loquacious, a full two minutes passed before Sherry uttered a word. "This is great," she said to Jen.

"Just unbelievable," Jen agreed.

A few feet away, at Table 24, Leslie Pooser was also experiencing feelings of disbelief. The former science and chemistry teacher, a thirty-seven-year-old single mother who commuted every day to Newark from the Brooklyn neighborhood where she'd lived her entire life, Leslie had assumed that issues dealing with death and dissection would be addressed prior to the onset of dissecting. Never did Leslie imagine that she and her classmates would "plunge right in."

In drawing up the lab assignments, Vasan had made a conscious decision to place Leslie at a table with Cary Idler. Another nontraditional student—the school's definition of any student over the age of twenty-five—Cary, twenty-nine, had spent his immediate postundergraduate years building and designing houses. Eventually, his passion for science and medicine landed him at NJMS as a research assistant. By the time the gross anatomy lab doors opened, Cary had a lot of experience dissecting. And though that experience was limited to laboratory animals, on the day lab began, Cary already bore the demeanor and confidence of a skilled surgeon. In temperament and skill, Cary Idler seemed the perfect person to shepherd Leslie Pooser through the first stage of the dissection.

On day one, not even Cary's presence could counteract the trauma experienced by Leslie Pooser. With each swipe of the scalpel, Leslie apologized. Not to her lab partners, but to the cadaver. As the afternoon progressed, the regrets transcended the laboratory: "Forgive me, God. Please, forgive me," she intoned, looking skyward with every cut.

In Lab D, Ann Waldman* encountered a different problem. Earlier, she'd been astonished when another student became faint as his lab partner lifted the towel from their cadaver's face. Soon after, the student returned, explaining that it hadn't been the sight of a dead body but a persistent case of stomach flu that sent him reeling. (While every medical student fears passing out, it rarely happens. A notable

*Not her real name.

exception was an incident thirty years ago in a Midwest anatomy lab that, to compensate for a shortage, had imported cadavers from other states. On the first day of lab, one of the students took one look at his cadaver and fainted dead away. The body had come to the school, as had he, from Texas: The body was that of a recently departed aunt who'd donated her body to science.).

It wasn't the sight of the cadaver or a touch of the flu that made Ann want to retch; it was the odor of the body atop Table 37.

From the moment of the first incision, it was obvious something, either the cadaver's metabolism or a chemical imbalance triggered by the embalming process, had prevented the phenol-glycerine from properly saturating the tissues.

Whatever the reason, the body on Table 37 was so pungent that Roger Faison was summoned to determine what could be done to rectify the situation. He recommended they air out the cadaver by not closing the hood for a few days.

By design, Nagaswami Vasan, a hands-on course coordinator who moved from room to room overseeing the instructors and students, tailored the first-day exercise to provide on-the-job training. The primary goal, reflecting the skin, acclimated the students to the movement of the scalpel and the key role of the forceps. ("They're tweezers to everyone else. But they're not tweezers to physicians and residents," Jen said, repeating a dictate caustically passed along to her during a hospital pharmaceutical internship.) Before the afternoon was over, the students were expected to cut through the primary and deep fascia and identify the pectoralis major muscle as well as the attendant circulatory and nerve systems. By four o'clock, two hours into the dissection, it became clear Table 26 would not meet the objective.

The dilatory pace was partly their own doing. Intent on making a favorable impression, the group went out of its way to ensure all had equal turns at dissecting, aided in the quest for equanimity by the fact that the chest provided ample space for two anatomists to work simultaneously.

Insisting that they stop every few minutes to review, Sherry good-naturedly accepted responsibility for the laggard pace. "What can I say? I'm older. I have to look at things a few more times than I did when I was in my twenties," she informed her lab partners. As long as it didn't intrude on their own cutting time, the others minded Sherry's deliberation not a bit.

Of the four, only Ivan hung back. Jen, sensing her teammate needed a boost, adopted the role of cheerleader. "Ivan, you're looking pretty clean over there," she said encouragingly as Ivan huddled over the cadaver with the scalpel.

"Yeah. Right. Unless I get tired. Then there will be chunks all over the place," he said.

"Once you begin to plane it's easier," Sherry suggested.

As if on cue, Ivan's forceps slipped, causing the scalpel to spray a glob of fat onto his face. "That'll teach me to keep my mouth open," he cracked.

"Now I know why I'm a vegetarian," Sherry said blandly.

In all her years as a nurse, Sherry had never seen anything like Number 3426. The human body, coursing with blood and oxygen, is vibrant. Even in duress, the organs, nerves, arteries and veins are easily delineated. Conversely, a body embalmed and then kept frozen for an extended period of time takes on the consistency of beef jerky.

In a way, the donor reminded Sherry of the accident victims she saw while working intensive care units. Fresh out of surgery, they arrived in bodies broken and twisted, their faces grotesquely swollen.

She hated to admit it, but until the families showed up teeming with questions no one could answer, it was often easy to forget they were human. In cases when the hourly vigils dragged into days, the reminders became tangible: snapshots of the injured taped to the wall over the bed, Mylar balloons, and other artifacts. The photos depicted men and women in the flush of life: golfing, waterskiing or enjoying a quiet moment in a favorite place, often in the company of a spouse, child or grandchild. Sherry could see how the snapshots brought a sense of comfort, if not an unspoken glimmer of hope, to the families. For the staff, professional detachment being the font of sanity, the snapshots had the opposite effect. Distracting and disconcerting, a photograph had a way of humanizing the process, making it even more difficult when the technology failed. For the staff, it was much easier not knowing who or what they were before.

Sherry was glad that no snapshot of Number 3426, as he lived his life, would ever appear above the table where he rested. And yet the thought that she wanted to know more not only about who he was but what, exactly, motivated him to donate his body kept tugging at her. Sherry wished for a reminder of his humanity; she wouldn't have to wish long.

One of the first indicators of the life led by Number 3426 was the amount of fatty tissue in his chest. Clearly, this had been a man who enjoyed a good meal. Large but not corpulent, Number 3426 nonetheless stood out in a room filled with bodies emaciated by cancer.

Matthew Gewirtz passed Table 26 en route to his locker and eyed the thickness of Number 3426's chest. "Gross," he said, turning up his nose. Matt, at Table 23, had a lean cadaver.

"Hey, don't be making fun of our guy," Jen admonished. Her tone was jocular, but she meant it. An athlete at heart, Jen prized the concept of teamwork; already she recognized Number 3426 as the most crucial component of her new team.

"Yeah, we like a little meat on our men," Sherry added.

By five o'clock, the designated end of lab time, only the students at Table 26 remained in Lab C. "I'm proud of our table. We're taking a long time, but we're doing it right," Jen boasted just as her scalpel snagged a barely visible white thread.

"Is that a vessel? Why can't I cut through it?" Jen asked Udele.

Having already quietly assumed the role of leader, Udele studied the structure. "Nerve," she said assuredly.

"Nerve? You sure? It seems really fat. Not that I've seen a lot of human nerves. How do you tell the difference between nerves and arteries?"

"Nerves have a lot more fiber," Udele said.

"They're stringier," Sherry added, focusing on the nerve clamped by Jen's forcep. "Lateral thoracic," she declared with a hint of uncertainty.

Jen picked up a scalpel and began scraping more tissue, working carefully, but not cautiously enough for Sherry, who gasped each time the blade passed within a millimeter of Jen's left thumb. "Sorry," Sherry apologized. "It's the mother in me."

Preoccupied with locating the lateral thoracic blood vessels, Sherry pulled out her *Netter's Atlas* and turned to Plate 175. Some had expressed surprise that Sherry had brought a $60.99 textbook into lab; Sherry believed sacrificing the book to the first-year smell would bring its own reward.

"There, that's it," she said excitedly, pointing to the *Atlas* and then to the sinewy structures pinched in Jen's forceps. "The lateral thoracic vein and artery. Oooh. Look at this vessel. How cool is this vessel?"

Still working on the opposite side of the table, Udele had her own reason to be pleased as she triumphantly lifted and displayed the pectoralis major. Demonstrating to Jen how to do the same, Udele's face was aglow.

Lifting the muscle, Jen sensed something amiss. While in *Netter's*, as in all atlases, hues of red and blue highlight the various structures, inside a cadaver tincture is generally nonexistent. Except for those bodies containing a well-preserved (green) gallbladder, the human body frozen for an extended period of time is of uniform color: an indistinguishable grayish brown. Yet as she peeked under the pectoralis major, Jen could have sworn she saw a swatch of blue.

"Just like the book," she said suspiciously. "I'd be surprised if it looked just like the book."

Sherry leaned in and authoritatively identified the aberration as sutures, the initial evidence to the four students that they were not the first to plumb the depths of Number 3426's thorax.

"There's a treasure in there," Jen predicted.

Sherry jumped as the pager in her pocket began vibrating, a reminder that her fourteen-year-old son would soon require a ride home from wrestling practice. "Gotta go," she said apologetically.

Before leaving, Sherry removed a flap of skin dangling uselessly under the cadaver's armpit. "I think it's OK if we cut this off," she reasoned. "He won't get cold tonight."

Sherry's departure allowed Ivan to bring up to Jen something he'd been dying to mention for the past hour. Afraid of embarrassing her in front of two people they barely knew, he'd held off. Now, with Sherry gone, he discreetly informed Jen about a small glob of fat stuck in her hair. Jen reached up and pulled it out. "Oh, God, Ivan. Why didn't you tell me before?" she said, sweeping it onto the table.

It was six o'clock and no one, save three of the four students assigned to Table 26, remained in any of the four labs. Before Udele placed the towel across the cadaver's chest, Jen sprayed him with water, bemoaning the end of a memorable afternoon. "This has been totally cool," she told Ivan.

As they closed the hood the conversation turned, briefly, to the evening's dinner menu at their respective apartments. Strangely, Ivan and Jen were famished.

"Chicken?" Ivan ventured.

"Don't think so," said Jen.

CHAPTER 5

D inner, courtesy of her roommate, Christine Ortiz, was on the table the minute Jen arrived home. Christine, a twenty-seven-year-old intensive care unit nurse turned medical student, had also departed lab mystifyingly hungry, and assuming Jen would be starving, too, she rushed back to the apartment they shared to immediately put water on the stove. With due consideration to what she and Jen had just experienced, Christine took the vegetarian route: pierogi and beans.

While Christine and Jen wolfed down dinner, Ivan went hungry. Somewhere between school and his apartment, the bottom dropped out of his appetite, a circumstance he blamed on double-gloving that hadn't achieved the desired effect. Every time Ivan brought his hands anywhere near his face, he caught a whiff of phenol.

For some reason, an odor that hadn't fazed him in lab began making him sick once he got home. When soap and water proved unsuccessful, Ivan hatched a plan to mask the smell by scrubbing the bathtub. For a few minutes, the ploy worked as the bleach-fresh scent of the cleanser overpowered the embalming solution. Within the hour, phenol again won out and, when it did, Ivan gave up and tumbled into bed, instantly falling asleep.

Sherry's olfactive issues prompted the kids to inform her she reeked, an assessment not taken personally. Considering a yellow surgical gown over street clothes sufficient for the purpose, of the 181 students in lab Sherry was the only one who hadn't worn scrubs. Nor could she understand why so many of her classmates had gone to the trouble of double-gloving: If nothing else, years of practical experience had taught her, aphoristically, that medicine stinks. To Sherry, the smell of phenol and decomposition were just two more odors encountered along the way.

At nine o'clock, after the kids were in bed, Sherry picked up her physiology textbook. To the dismay of everyone in the class, NJMS had paired in the same semester the two bedrocks of medical education, gross anatomy and physiology. The students were not alone in registering disbelief: Even the faculty found it difficult to fathom that their charges were expected to simultaneously assimilate the complexities of two of medical school's toughest courses. The competition between the anatomy and physiology departments was understated but fierce, and in the battle for the minds of the Class of 2002, physiology flexed its academic muscle most often by scheduling an exam for the second week of the semester and every two weeks after that.

Sherry spent ninety minutes on physiology; then, around 10:30, Richard S. Snell's *Clinical Anatomy for Medical Students* came out of the backpack. Fearful of the consequences should she fall behind, she began reviewing the clinical notes pertaining to the next day's dissection. After two pages, her eyes grew heavy; the third page she left unread.

Before Jen, the jock, nodded off, she noticed her arms felt unusually heavy and sore, the way they used to after a particularly rigorous preseason basketball workout. She racked her brain trying to determine the source of the deep muscle pain. The only thing she could come up with was the repetitive motion of the scalpel.

Udele, as usual, ate dinner in the cafeteria and studied in the library. When she arrived home after midnight, the rest of the family had long been fast asleep. Soon, so was Udele.

Not everyone in the Class of 2002 found detachment and repose as easily as the team at Table 26.

The moment Ann Waldman walked into her apartment she collapsed, catatonic, into an easy chair, where she spent the night staring at the wall, rejecting a relentless effort by her live-in boyfriend to discuss the events of the day.

When Leslie Pooser returned to Brooklyn, her ten-year-old son, Omari, was also anxious for details. Normally, Leslie and Omari shared every part of their day with each other. That evening, all Leslie wanted was to vegetate. She didn't want to study, she didn't want to read, she didn't want to interact with anyone. All she wanted was to sit quietly and hope that, somehow, weariness would turn to sleep. Omari kept pestering, so, to appease him, Leslie offered a brief description about the color, consistency and odor of human fat. With that said, she swore she'd never again eat barbecue ribs.

On the morning of August 17, 1998, Leslie had entered what the New Jersey Medical School calls the Great Hall and what architecturally qualifies as a glorified atrium. She located the table alphabetized for students with last names beginning with the letters M through R. From a file box, an admissions clerk removed an envelope bursting with forms and other minutiae deemed important by the NJMS bureaucracy. Leslie's attention was not on the myriad paperwork being thrust at her but on the name typed boldly across the top of the information packet: LESLIE POOSER.

Oh, shit, Leslie thought. This isn't a joke, this isn't April Fools' Day. This is real. Oh, my God. They're expecting me.

She had wanted this moment from the age of seven, and everything subsequent to that—even obtaining an education degree from the City College of New York—became a means to this end. While pursuing her degree, she stayed abreast of the latest developments in science and medicine by working part-time at jobs that brought her into contact with physicians, an exposure that led Leslie to classify half of all doctors as "idiots."

In med school, she came to understand why intelligence and medicine are not mutually inclusive: "It's about repetition and memorization. You need to know when to pick up a book and where to look in it once you've picked it up. That's what research is, isn't it? It's knowing that when someone comes to you with a tingling in their arm that it's a problem with the brachial plexus. And if you're not exactly sure that's the problem then you go somewhere in a book to find out what can be done about it."

Leslie considered teaching a prelude, a way to earn and save money for what truly mattered: medical school. Fortunately, her love of teaching chemistry and science to other people's kids kept her intellectually focused on the primary goal. Leslie taught evenings, summers; in fact, she was in a classroom until three days before reporting to NJMS.

"That day—orientation—I knew. I knew it had finally come. I've been on a high ever since. I treasure every day of this because this is what I've wanted for thirty years," she said.

For the whole of the first semester, cell and tissue biology and biochemistry couldn't knock Leslie off that high. But the first days of winter term came close. With the exception of the mandatory visit to the gross anatomy lab during the NJMS preenrollment tour, Leslie's encounters with mortality were of the norm: wakes, funerals, nothing traumatic or extraordinary.

When Leslie Pooser's life planted her squarely in the midst of death, she told herself, "We need to do this, I know that. This brings everything together for me." Sensing how profoundly she differed from Sherry and Udele, Cary Idler, Jen and others who managed to cut through skin with apologies to neither cadaver or a deity, Leslie attributed her classmates' apathy, feigned or otherwise, to either prior exposure to clinical experiences or "bottled up" emotion. Not one to stifle her own feelings, Leslie confessed she had come to lab emotionally unprepared for what lay ahead.

In all candor, Leslie differed from many of her classmates in only one regard: the willingness to acknowledge apprehension.

Attempting to probe the repressed emotions of medical students, Dr. Sandra Bertman, in the early 1990s, began inserting a blank piece of paper into the orientation packet sent incoming students at the University of Massachusetts Medical School. The professor and director of the school's Department of Medical Humanities asked the students to transform the paper into a canvas upon which to compose an image reflecting their perception, from a purely emotional standpoint, of the coming dissection. As artists, she found medical students made pretty good physicians. Unconcerned with aesthetics, Bertman maintained her focus on content and, in that, the commonality in the drawings and accompanying text rarely disappointed.

Hundreds of sketches, each accompanied by written explanation, have come across Bertman's desk since the inception of the program. The images are stark and arresting, none more so than the depiction of a cadaver flanked on one side by the devil and on the other by an angel. The devil holds the tools of dissection, the angel a cup labeled "knowledge." "The person on the left benefits from the work on the right. Hopefully, the future will bind the two when knowledge is used to help others," the student explained.

If there exists a dominant theme, it is that of reverence: the image of lab partners standing over the cadaver, their hands, as well as those of the cadaver, clasped in prayer; a cross, in the form of a headstone, placed over a copy of *Gray's Anatomy*. Students who find art to be an insufficient form to express their feelings about dissection sometimes invoke musical parables, notably Mozart's *Requiem*.

Bertman summarized the experience in a quote published on the cover of a compilation of drawings, poems and essays: "It is commonly known that medical students dissect the bodies of the dead, it is less commonly realized that these same dead do a great deal of cutting, probing, and pulling at the minds of their youthful dissectors."

And as any medical student will testify, when the subconscious tugs, it does so mostly at night.

For one student, the passage through the NJMS gross anatomy lab a few years prior to the arrival of the Class of 2002 was fraught with anxiety brought by a simple reality: He didn't like to cut. "The stress kills you," said the student. "You don't want to cut through [the wrong] nerve. You keep thinking, 'What if it's a real person?'" Appreciating the importance of maintaining the proper outward demeanor—blasé and jocular—in lab, the student kept his misgivings in check.

The price he paid for keeping impulse in abeyance came the moment his head hit the pillow and the image of the cadaver, sliced open and sometimes literally pulled apart, loomed before him. While he could do nothing to control them, the student well understood the source of episodes that continued until the midpoint of the semester: "It's in my subconscious, I'm relating to how we cut. I'm relating to how we're cutting into a human being. Before I go to sleep at night the cadaver is on my mind. I bring the stench to my bed," he said shortly after the bouts of anxiety ended.

At least the student didn't dream the dreams or, if he did, he blessedly couldn't recall them.

Prior to lab, Ann Waldman had never before seen a dead human body. Twenty-two years old and freshly graduated from Syracuse University, she'd been touched by death only once. The previous year, her grandfather succumbed after a long and horrific illness. Five years from the time of diagnosis, the pain became so agonizing that, as the end drew near, he forbade visits from his beloved grandchildren. Afterward, in accordance with Jewish observance, the casket had been closed.

So, like Leslie Pooser, Ann Waldman first witnessed death in the flesh inside a gross anatomy laboratory, an environment that somehow conspires to make it quite easy to forget what is omnipresent there. In the room designated B527 by the New Jersey Medical School, death is devoid of drama: gone are the pretenses of stage and cinema, where the cessation of life is accompanied by the swell of a violin or a cacophony equal to the violence being portrayed. In a dissection lab, the bona fide human responses to death—grief, sorrow, anger and regret—are stripped away. Death, in effect, becomes banal.

For some students, the passage of death from grim to mundane occurred seamlessly, a discomfort level that evaporated without notice. "A lot of people had ideas about what to expect; not me," said Ann, who during the first week confronted a number of problems, not the least being assigned to the table with the cadaver so odoriferous that, when they weren't cutting, Ann and her team retreated to the nearest corner of the lab to escape the smell.

By the end of her second day in their company, Ann realized her lab partners were also going to pose a problem: Not only did Ann find them insensitive to her adjustment problems, she resented how the rest of the table rushed through the dissection. Ann, like Sherry Ikalowych, learned best in an academic atmosphere that allowed her to learn by taking her time. If the trade-off for identifying and memorizing key structures was putting up with the smell for a few extra minutes, Ann was willing to live with the odor.

Though she'd come to lab emotionally unprepared, Ann was pleased at how quickly she'd become comfortable using the scalpel as a cutting mechanism and the forceps as a tool to scrape away extraneous tissue. She was much less at ease with her lab partners' habit of using their hands to rip the skin and snap the nerves and arteries.

"The way we're abusing the body, pulling out the ribs and stuff, is that respecting him?" Ann posed the question to Jen over lunch; four days had elapsed since the first cut. "It seems abusive. I feel like we're hurting him. We just tear through it, like a chicken. Except, I wouldn't tear up a chicken like this."

"That's part of it," Jen said, postulating that violating the body, often in ways that violate statutory common law, is part of being a physician. "What about when you'll have to do a rectal [exam]?"

"That's different. The person is alive."

"But this is what [the donors] wanted us to do. They expected it. These bodies are donated. They weren't picked up off the street," Jen countered, unable to elevate Ann from her funk, understanding that at the core of Ann's unrest were the clumps of ball-bearing-sized black nodules inside her cadaver's thoracic cavity. On the list posted on the stanchion at the front of Lab D, the cause of death for the donor on Ann's table was stated matter-of-factly, in cold clinical terms: melanoma—the same disease that had killed Ann Waldman's grandfather not twelve months before.

Ann already assumed that one day cancer would kill her as well. The history was certainly there: In 1954, the disease had taken the life of her mother's brother at the age of two. Then, six months after her grandfather's death, a routine mammogram revealed her mother's breast cancer. By the time Ann moved from Syracuse to Newark to begin med school, a protocol of intensive radiation and chemotherapy had pummeled her mother's cancer into remission.

Ann arrived in Newark prepared for the long haul as a postdoctoral candidate, meaning that simultaneous to attaining a medical degree she would be working toward a Ph.D. She figured the process, including medical internship and residency, would take twelve years. With degrees in hand, Ann planned to apply the Ph.D. toward research and the M.D. toward healing: Ann Waldman wanted to become an oncologist.

The path traveled directly through Table 37, and while academically Ann hadn't the slightest doubt she'd cruise through GA, part of her worried that the emotional element of the dissection had the capacity to impede her path through science and medicine.

The recurring dream that exacerbated her fear was never about cutting or medical students using their hands to snap nerves and arteries. The cadaver never made an appearance, Ann said, relating the content of the dream to Jen in the cafeteria. It involved carnage and randomly grotesque street violence, its victims children.

"I haven't had a dream yet," Jen confessed with a shudder, hoping that would continue to be the case.

"You're probably having them. You just don't remember," said Ann.

An unpublished study conducted by anatomy instructor Dr. Anthony Boccabella found that while most students don't recall their dreams, for some, a torrent of anatomy-related dreams during the onset of dissection drops off dramatically as the class progresses. As students grew accustomed to the procedure, Boccabella found, so did the subconscious, until, two or three weeks into the semester, the dreams about death and bodies and desecration dissipated, replaced by signature student nightmares about sleeping through or botching an exam.

That Boccabella bothered to examine the psychological impact on students during the 1970s was considered unusual at the time. Until only recently, medical schools were pretty much content to throw students into human anatomy labs without consideration to the short- and long-term emotional consequences. Along with a colleague, UMass's Sandra Bertman wrote in a 1985 study that "Traditionally, students have seldom been encouraged to voice [problems adjusting to anatomy courses]; their attention, and that of their teachers, has been focused on the mastery of the curricular material at hand."

To illustrate the long-term impact, Bertman cites *A Parting Gift*, the 1982 medical school chronicle in which a physician attributes her later "helplessness in facing death" to the "atmosphere and experiences of the dissection laboratory."

Medical educators consider Bertman a pioneer, for until the art historian and former teacher began analyzing the relationships between students and cadavers as part of her landmark studies on death and dying, the emotional component of gross anatomy was generally ignored.

A prominent observer of modern medical education and practice considers that a colossal oversight. "Of all the professions, medicine is one of the most likely to attract people with high personal anxieties about dying," wrote Dr. Sherwin B. Nuland in *How We Die.* "We became doctors because our ability to cure gives us power over the death of which we are so afraid. The loss of power poses such a significant threat that we must turn away from it."

Certainly, no one dealt with emotional issues when Dr. Rita Charon matriculated at the country's most prestigious medical school. At Harvard in the mid-1970s, Charon perceived dissection to be an almost furtive endeavor, recalling it as a "very hidden, almost prohibitive activity that we were taking part in." By example, Charon, now an assistant professor of clinical medicine at Columbia University's College of Physicians and Surgeons, cited the afternoon she spotted a pair of university electricians observing the proceedings through an open anatomy lab door: "I ran and closed the door because I thought they shouldn't see what we were doing. So, there is certainly a feeling of doing something disallowed and I think that is rather common."

Never, not during the course or in its aftermath, did Harvard do anything to dissuade that notion. Back then, the idea that a medical school might hold a memorial service to honor the men and women who donated their bodies was unheard of. Today, Charon said, "I think schools are being more and more responsible by acknowledging the personal trauma inherent in this exercise and by devising different ways to address the students' emotional and spiritual reaction."

Still, because schools generally wait until after the fact to address the issues raised by dissection, students are pretty much expected to deal on their own with whatever trauma they might experience along the way.

If there is an alternative to simply throwing students into the mix, Dr. John H. Siegel, the chairman of the NJMS Department of Anatomy, Cell Biology and Injury Sciences, doesn't know what it might be. Somewhat puzzled by adjustment problems, Siegel pointed out that it can be reasonably expected that the student who has charted a career in medicine is fully aware of what he or she is getting into before arriving at medical school. At the same time, Siegel acknowledged that even the most intense personal preparation may not be enough. "It is a brutal initiation, no doubt about it," the chairman admitted.

"A messy domain" is how Charon put it. Yet, like Siegel, the Columbia professor remains unconvinced that anything can or should be done to lessen the impact: "I wouldn't put it as a requirement to anatomists that they spend curriculum time teaching the psychology of human sympathy. Certainly, they're obliged not to diminish or disrespect the dead. But with the kind of efforts that anatomy courses are now going through to prepare students and allow them to talk about feelings and what with the memorial services and writing about the bodies after the fact, these are really very extraordinary ways of being responsible."

Charon and Siegel have history on their side since, for all the flaws and indignities medical institutions inflict upon students, there is little to suggest the system does not work. Somehow, each year, all but a handful of the med students who walk through the doors of human anatomy labs stick it out for the entire semester.

For the squeamish, passage through anatomy requires a summoning of fortitude and whatever other means are necessary to overcome fear and revulsion of the assignment, even if it means donning a surgical mask to ward off the smell, as was the case for a handful of NJMS students of the Class of 2002 as it moved toward the final and most critical procedure of the first week: the removal of the lungs.

As Table 26 plodded, sans masks, toward the goal (indeed, by the second week the masks had disappeared entirely from the lab), they maintained a pace that accommodated learning to the exclusion of efficiency.

Last among equals, by the time Ivan Gonzalez's opportunity to wield scalpel and forceps arrived, the pertinent structures in the tissue surrounding the lungs had been long uncovered. Not once during the first three days of dissecting did Ivan reveal an artery, nerve, vein or any other configuration to cause Jen to exult. And Jen exulted often: Minor discoveries earned a *cool*; major revelations a *wow*.

Unsure of his own ability with the scalpel, Ivan didn't mind being relegated to a supporting role. From the start, he thought it only natural that Udele be awarded the number-one ranking, and given her experience, Sherry certainly had earned a prominent place in the pecking order. But in terms of experience, Ivan knew he and Jen were pretty much equal. While he accepted that a certain hierarchy would emerge, he was surprised it had transpired so quickly. He never imagined that before the end of the first week he would be, by virtue of gender, the low—and lone—man on the Table 26 totem.

His station became clear on the second day, when Sherry dispatched him to the tool room. "The man does the sawing," she said, establishing a precedent. From that moment forward, every task beyond scalpel and forceps dissection fell to Ivan. He'd come to lab expecting to learn many lessons; finding out that sexism works both ways was not one of them.

Unwilling to upset the emerging group dynamic, Ivan resisted the temptation to lash out. As an undergraduate, he'd been the one to reach out to classmates in need of emotional and academic support, a trend that continued during the first semester at NJMS, when he tutored his classmates in Spanish. And though he put on a good front, privately Ivan felt increasingly isolated and, much to his chagrin, self-centered.

Cognizant that a blunt retort to Sherry's directive would further reflect an emerging side of himself he didn't much care for, Ivan held his tongue and dutifully headed for the tool room to borrow the tools critical to cutting through the ribs in order to extract the lungs: a stainless-steel hacksaw and a pair of vicious-looking shears, the stainless-steel blades ominously curved in the pattern of a hawk's beak.

Before they reached the lungs, Table 26 had first to navigate an impossibly difficult anatomical obstacle course. Two maps illuminated the journey: *Grant's Dissector* guided them through the physical landmarks and told them what they should know; Vasan's course syllabus provided the academic context, with information imperative to what they had to memorize for the first exam.

Never mind the multitude of serried nerves or the perforating and collateral branches of the primary blood vessels, the vast number of major arteries and veins in the thoracic cavity staggered even Udele, who'd been there before. The two days of shredding by scalpels and forceps had transformed the cadaver's relatively pristine, though surgically scarred, thoracic cavity into an indecipherable morass.

The superior epigastric and musculophrenic arteries, the intercostal artery and vein, the pulmonary artery and vein, the list of what they were expected to identify within the havoc Table 26 had created was endless. Worse, Jen couldn't even pronounce half of the stuff. "Don't worry about it," Ivan assured her. "Just know where it is."

Equally overwhelmed by the volume of information, Sherry futilely searched Thursday afternoon for the internal thoracic artery—a vessel the syllabus had deemed important. Time was growing short: By tradition, Thursday was the last day lab met formally before the weekend. Finally, she gave up and raised her hand, the med student sign of surrender.

Others were also capitulating. At any given time on any given day, at least half of the tables in Lab C had their hands in the air, beckoning who now was the lab's sole instructor, Zolton Spolarics, to answer the room's infinite questions about the anatomical maze. Tall and bald with penetrating blue eyes, the intimidating perception Spolarics projected to the uninitiated was further proof of how looks deceive. Soft-spoken and a patient listener, Spolarics possessed the trait inherent in all good teachers: unlimited patience.

"The internal thoracic, we can't find it," Sherry said when Spolarics finally strolled up to the table.

Sherry placed the forceps in the palm of Spolarics's hand and the instructor went right to work, knowing instinctively what he was looking for and where to find it. Thirty seconds of peeking under and around thoracic tissue told him all he needed to know.

"That's because it's not here," he said.

Sherry appeared stricken. "We cut through it?" she asked incredulously.

Spolarics nodded solemnly.

"And we need it, right?" Sherry continued, already thinking ahead to the practical exam, the hour-long ordeal during which the students were expected to identify structures tagged by the faculty.

Spolarics again nodded, this time adding a reassuring smile. "Don't worry, you'll find it on one of the others," he said, pointing to Table 25.

Turmoil identical to that at Table 26 unfolded throughout the lab as structures, key and otherwise, were obliterated at a rate that sent waves of panic through the ranks of the students. The epidemic of destruction in the thoracic cavity is something course coordinator Nagaswami Vasan, with nearly a quarter century spent in anatomy teaching labs, fully expected and, in a way, hoped for. Over the years, as Vasan witnessed both students and faculty come and go, the script had remained the same:

"The first week, they don't know anything. It's all isolated information. Then, in the second week, it starts to come together. By the third week, it starts to connect. There are enough connections that they can begin to forecast the future. All of a sudden, they know that there is something in front of a nerve or muscle, that there is something behind it. They begin to work with a certain caution, because they know what's coming later on."

Circumspection had yet to enter the equation when Ivan returned to Table 26 with shears and hacksaw. Undertaking what Sherry referred to as his "manly duty," he proceeded to sever the ribs with the shears, the first step in freeing the breastplate covering the lungs and heart. The sound of metal crunching through soft bone and marrow resembled a paper bag crinkled loudly. Sherry grimaced; Ivan looked pleased: For once the center of the table's attention, it felt good to contribute.

"He's leaking! Nurse! Do something!" Ivan exclaimed as a grayish-orange liquid seeped from underneath the rib, Table 26's first encounter with the anatomy lab by-product known as cadaver juice. Sherry, who thought she'd seen everything that medicine had to offer, had never seen anything like the result of embalming fluid combined with the liquid generated by thawed human tissue. Jen took one look and went off in search of one of the lab's two aspirators, known colloquially as "juice machines," her departure grinding to a halt Ivan's moment in the limelight.

As soon as the liquid was suctioned out, Ivan returned to the task. To give him better access, Udele and Sherry spread open the cadaver's arms, creating a Christ-like effect.

Ivan moved to the right side of the body and Sherry manipulated the severed ribs on the left, pointing out, "If you hit the wheel of a car, this is essentially what you do to yourself."

The bluish sutures visible the first day through the thin layer of tissue immediately under the skin were, in fact, staples, five in all, placed approximately an inch apart. As Ivan sawed through the sternoclavicular joint, the saw kept glancing off the topmost staple. Exasperated, Ivan paused just as another person joined the three lab partners witnessing Ivan's performance: Vasan, in the midst of his daily lab rounds.

Vasan's presence was daunting and his stature such that when he visited one table, students from other tables invariably filtered over to perhaps snatch a nugget of wisdom. Ivan hoped this time Vasan wouldn't draw a crowd.

"How do we know if we're going too deep?" he asked.

"The guy will scream," said Vasan.

The saw broke through, but Ivan wasn't finished, as he hit another snag: The sutures and adhesions created by Number 3426's surgical encounter necessitated exorcising a full rib in order to facilitate the removal of the breastplate. Without a word of instruction from Vasan or his lab partners, Ivan picked up the beaked shears and went to work.

While he cut through the rib, Jen hovered over his shoulder, adopting the role of medical commentator. "This was not a good area for him. A lot went on up here. Poor guy, he has a ton of scar tissue. I can only imagine all the medications he was on," she said.

Ivan stopped cutting for a moment and rolled his eyes in her direction. "Pharm major," he mocked. Ivan tugged at the rib; it broke off cleanly, prompting Sherry to remove, as though it were an external protective armament, the breastplate. Setting it on the towels covering the cadaver's legs, she immediately returned to see what secrets lay beneath in the now fully exposed chest. "The parietal layer. How cool," she marveled.

Because word travels fast in a human anatomy laboratory, news that Table 26 had an anomalous thoracic cavity ricocheted around Lab C like a pinball. Within a matter of minutes, fifteen students surrounded Table 26 as Udele, Sherry, Ivan and Jen, beaming like proud new parents, showed off the sutures in the sternum, promising their classmates more revelations in the very near future.

As everyone drifted back to their tables, Sherry set out to make good the pledge, cutting through the parietal layer, the elastic tissue protecting the lungs, confident that underneath lay clues to the man's medical history. Number 3426 did not disappoint. In the snarl of veins and arteries, Sherry immediately detected an inconsistency: Another section of the internal thoracic artery appeared to be missing, and this time, it appeared to Sherry, it wasn't her fault. She raised her hand; Spolarics responded immediately.

Sherry suggested that the missing artery would reveal itself during inspection of the heart. Spolarics agreed and added that if Number 3426 had bypass surgery, the replacement vessel could just as well have come from another part of the body. Right on cue, Udele lifted the towel covering the cadaver's lower leg. She paused, studied and then, her face aglow, triumphantly pointed to a small scar between Number 3426's left shin and calf: an incision scar.

"I'll bet he had two of them," said Sherry. Spotting Jen's quizzical look, she explained, "Two bypasses; in the eighties, when they first started doing them, they used to take the veins out of the leg. The veins didn't hold up. A lot of guys who had bypasses then had to go back later to have them redone using arteries and stuff from the chest."

The offhand dissertation sent Jen reeling: It was precisely what she'd hoped for three days earlier when she saw her name next to Sherry's on the table assignments.

Forever making a connection between anatomy and the clinical functions of medicine and the body, Vasan often tells students that inside each cadaver there exists an individual drama, pathological testimony as to how someone lived his or her life. Plaque-clogged arteries in the abdomen suggest a red-meat eater; black-encrusted lungs signify long-term nicotine addiction; the faint indentation in the ulna is the scarred evidence of an arm broken in childhood.

Once, during the dissection of the pelvic region in the pre-Viagra era, the titter of nervous laughter drew Vasan to a table in Lab D. As he approached, the level of mirth dropped precipitously and the gathered crowd quickly dispersed. The source of entertainment was a fully operational penile implant pump. Vasan summoned the students back to the table and gave the pump a healthy squeeze. In gross anatomy, a cadaver with an erection is the ultimate sight gag. The students cracked up, as did Vasan. While treating cadavers with respect and dignity is always the top priority, humor, properly administered, is the best defense mechanism in dealing with what occurs in lab day after day.

Vasan turned serious and asked the students to consider the circumstances that led to the implantation, a process that undoubtedly began with humiliation and reached a point of acceptance. Resolution required a visit to a urologist and possibly more humiliation from the experience of discussing the situation with a stranger. Finally there came the decision to proceed with the surgery. Did truth accompany the announcement to friends and family about the pending visit to the operating theater? Or was it passed off as a procedure to rectify a problem with the gallbladder?

Vasan "explained to them how the pump was implanted and how it was used and how it helped to change that man's life and what it meant to him as a man and as a person. And I told them that understanding what had occurred would help them appreciate how much a doctor can do for his or her patients."

After the penis had been dissected, Vasan returned to the table and retrieved the pump. Later, in his G-level office, he placed it in a shoe box along with a growing collection of pacemakers and other manufactured artifacts found inside the cadavers, constant reminders to Vasan of humankind's capacity to prolong life while easing its passage.

Born to an affluent Brahmin family of nine headed by a university administrator, Nagaswami Vasan's fascination with anatomy began early. His childhood home, in the Indian state of Madras, had an outdoor atrium that occasionally became the final destination for birds of flight. Vasan's four brothers and two sisters did what came naturally to children who come across a dead bird and conducted burial ceremonies. With Vasan, burial was a ritual undertaken only after he'd used a razor to perform a dissection. "I just always wanted to see what was inside," he explained.

In Madras in the 1940s, home delivery of milk meant bringing a cow directly to the doorstep. Striking a friendship with his family's milkman, the young Vasan enjoined him in conversations about lactation. Inevitably, given his predilection, Vasan's discussions with the daily visitor to his home led to inquiries about bovine anatomy. Not content to get information secondhand, the badgering continued until the milkman agreed to allow Vasan to attend a birthing. Simply bearing witness to the process was not enough for Vasan, who eagerly joined in, plunging his arm elbow-deep into the breach, pulling the calves into the world.

Calves with the misfortune to enter that particular part of the world as bulls normally didn't fare too well. In a Hindu nation, where cows are deified and religious law prohibits beef consumption, the bulls are considered an unnecessary extravagance, with only a minimal number required for breeding or plowing fields.

With the exception of a few animals put out for stud, a bull is valued for only his hide, the result being that most males never make it out of the barn. The mode of execution is quick and humane: The milkman incises the bull from sternum to abdomen with a single knife stroke. After the removal of the gut and muscle, the bull is stuffed with straw and placed near his nursing mother, a decoy to guarantee lactation. A good number of hardened cattlemen were known to recuse themselves from the proceedings with the approach of the evisceration. Not Vasan, who, at first opportunity, absorbed every step of the procedure. The next time he joined the milkman in the barn, he participated.

As he moved into his teen years, Vasan's obsession with anatomy grew ever stronger. On the evening his father brought home for dinner a colleague, an anatomist at the Madras University School of Medicine, Vasan stayed up half the night asking questions until, exhausted, the man finally had to beg his leave.

Given the perspective from which he'd always viewed humans and animals, when the time came for Vasan to choose a profession it seemed only natural that he gravitate toward medicine. Fearful that acceptance into Madras University might suggest nepotism, he instead adopted the opposite tack, rejecting medical school to obtain a degree in veterinary medicine.

As a practicing veterinarian, Vasan was something of a slacker, dedicating more attention to improving his backgammon and tennis games than he did to his practice. Even so, among his colleagues, Vasan quickly gained a reputation for fearlessly performing a most wretched task: cleansing the pus and maggots from open animal wounds—under no circumstance an activity engaged in by a Brahmin. The Indian caste system dictated that the function be left to "technicians," a subclass of the "untouchables." Having no use for the class distinctions that interfered with his preoccupation, Vasan pulled on "the gloves, pushed the technicians aside and did it myself. I was hands on. I just liked to see what was in there."

Veterinary medicine, which requires of its practitioners anatomical expertise for equines, sheep and goats, felines, canines and aviaries, served Vasan well when he eventually converted to a form, *Homo sapien,* that comes in one basic model. The scholarly path that brought the conversion began in Madras, passed through Denver and Philadelphia before ending in Newark. Along the way, Vasan picked up degrees in animal genetics and nutrition and a Ph.D. in neurochemistry.

Always immaculate in a freshly pressed shirt and slacks and designer tie, Vasan kept his salt-and-pepper pompadour perfectly coiffed. A firm believer that clothing with natural fibers retards odor, Vasan never changed before lab; a white lab coat, worn as a smock, served as the only protection for his clothes from the first-year smell and untidy laboratory splatter. As often as not, he walked around drinking from a can of Sprite, a practice that disgusted the students until they, too, began bringing soft drinks and coffee to lab. When they did, Vasan would regale them with stories about how students used to eat their lunch in lab when he taught at the University of Pennsylvania. No one believed it.

While Vasan may have been buttoned down sartorially, in every other way he brought an atypical approach to medical education, striving to be a friend as well as a teacher. Rare was the med school party he didn't attend. An accomplished chef, twice and sometimes more every year he invited groups of students to his home to sample homemade Indian cuisine. In an atmosphere populated by professors and instructors who steadfastly avoided eye contact, Vasan was a politician working the crowd, greeting students new and old by name, remembering details of their lives that most of the faculty had never bothered to learn in the first place.

Not everyone venerated Vasan: Resentful of his absolute control over every aspect of the anatomy curriculum—Vasan sat in the back of the room for every lecture and, in effect, graded his colleague's performances—the faculty bitched about him incessantly behind his back. And many students thought him to be imperious and condescending. "He's a prick. He tries to get on everybody's good side and make like he's friends with everybody. He's not. If you get on his bad side, you're screwed," one student complained over lunch one day. The burst of vitriol caused the jaw of the student's lunch partner to drop in disbelief; later, after his friend's departure, he attributed his surprise to having never before heard anyone utter the slightest word of criticism about the course coordinator.

Certainly, no one at Table 26 had anything bad to say about the Vasan who, to the team's chagrin, visited only once a day and was nowhere to be seen when they reached their first milestone.

After three days of cutting through skin, fat, sutures and adhesions, late Thursday afternoon Jen turned to page nineteen in *Grant's* and read: "Place your hand into the pleural cavity between lung and mediastinum. With one hand push the lung laterally, thereby stretching and exposing the root of the lung. With a scalpel in the other hand, carefully transect the root in the middle between lung and mediastinum. . . . Remove both lungs. . . ."

On opposite sides of the table Jen and Ivan cut while Sherry and Udele pulled, lifting the lungs simultaneously from the thoracic cavity. When the organs were placed atop the cadaver's abdomen the left lung, as *Grant's* had promised, was smaller than the right.

As they admired their handiwork, Jen used her finger as a pointer and began recounting the structures. "Bronchus, hilum . . . superior lobe . . . middle lobe . . . inferior lobe. This is great. I'm so impressed. It looks exactly like I pictured it."

Sherry—"I have a mind like a sieve"—retreated to the corner table to furiously scribble notes.

The removal of the lungs caused a veil to lift. Throughout the lab the sober concentration that preceded and followed the first cut absented itself. The laughter that now filled the room was genuine, nothing like the nervous twittering of two days prior, a smug satisfaction punctuated by exhilaration. They had done it, all of them: They had cut; they had dissected; they had survived.

An informal pilgrimage got under way, students circulating from table to table to examine lungs. "Cancer at twelve," someone announced. And everyone scurried into Lab B to witness the ravages of tobacco. "Surgery, half a lung at forty-five," came from Lab D, causing the entourage to head for that room.

Table 26, too, became a primary destination. The lungs resting on the towels covering Number 3426's abdomen were reddish-gray with white spots. Twice the size of the lungs at any of the other tables, each weighed six pounds.

"Sputum, from smoking," Sherry said as the steady stream of classmates trickled to Table 26, where Jen proudly stood sentry over Number 3426, her right hand absently patting his shoulder.

Udele picked up the lungs and motioned for the lab partners to follow her into Lab A to conduct the second portion of the ritual, a pleural tap. Brandishing a small hose, Udele carefully inserted the needle at the tip into the bronchus of the left lung and switched on the air supply. The slight inflation of the lung was quickly superseded by a large clear membrane attached to it that ballooned almost to the point of bursting.

"What . . . is . . . that?" Jen said in measured tones.

Udele opened the air valve two more times. "I don't know," she shrugged.

"A bleb," said Sherry.

"Bleb?" Udele and Jen sounded the word in unison.

"Bleb. An abnormality. Could have been from the disease that killed him," said Sherry.

As a unit, the four lab partners returned to Table 26. Passing Table 24, Sherry noticed a familiar scent.

"Air freshener." Leslie Pooser smiled, pleased with her addition to the spray water bottle.

"Good idea," Sherry said, unsuccessfully suppressing a yawn. "This is exhausting. Until this semester, we were sitting on our butts in lecture hall or in front of a microscope. This is much more physical."

Jen and Udele began packing up Number 3426 for the weekend by inserting the lungs into the thoracic cavity. Before covering the organs with the breastplate, Udele searched for a missing component. Under the cadaver's arm, she found it: the severed rib, which she inserted in its rightful place. Along with Sherry, she covered the cavity with the reflected flaps of skin while Jen went off to borrow Leslie's scented water.

As she closed the hood, Jen thought about how, so far, med school, with its insidious lectures and microscopic laboratory assignments, had not appreciably differed from pharm school. Today, however, there had been an almost imperceptible shift. Jen still felt like a student; there were too many lectures for that not to happen. At the same time, though, part of her was also starting to feel like something else. She couldn't quite put her finger on it, but she suspected it had something to do with the first stirrings of what it felt like to be a doctor.

CHAPTER 6

———————•———————

A lone but for forty-five dead bodies, two nights later Udele spent an hour in lab reversing her final act of Friday afternoon by retracting the skin flaps and lifting the cadaver's breastplate, a prelude to examining the lungs and contents of the thoracic cavity. Despite the semester being only a week old, Udele recognized that the preparatory process could never begin too soon and that, although the lungs were about to be obscured by the dissection of the heart, the pulmonary system would be accorded equal importance on the first exam.

The only member of the class that weekend to avail herself of an open-door policy that made the lab available twenty-four hours a day, seven days a week, Udele also had the advantage of proximity: She lived only ten blocks away.

For obvious reasons, no one in the Class of 2002 knew Newark as Udele did. In fact, by the end of the first week of lab, the Class of 2002 had already learned more about the human body than they would ever know about the city where many of them had lived and attended school for the past five months.

In its singular attempt to acquaint the students with the nation's third oldest city (after Boston and New York), NJMS had arranged during orientation an hour-long bus tour conducted by a historian who boasted of Newark's handsome heritage in architecture and urban planning along with the burgeoning renaissance represented by the New Jersey Performing Arts Center, a world-class cultural venue that opened on the edge of downtown in 1997.

As the bus rolled along its route, those passengers who weren't asleep took note of the grandeur incorporated into the design of the Essex County Courthouse by Cass Gilbert, also the architect of the U.S. Supreme Court. At Bound Branch Park, designed by eminent planner Frederick Law Olmsted, the impresario of New York's Central Park, they caught a glimpse of the country's largest grove of cherry trees.

They learned Thomas Alva Edison had lived and worked in Newark; it was there that he invented the ticker tape. And they were informed that the city is divided into five contiguous wards. Before the migration to the suburbs began in earnest in the 1960s, each represented an ethnic fiefdom: Italians in the North Ward; Portuguese in Ironbound; Slavics in the East Ward and, famously, due to the fiction of Philip Roth, Jews in the North Ward community known as Weequahic. The Central Ward, site of the New Jersey Medical School, was promoted as a center of commerce, education and government, not as a locale steeped in wrenching poverty and crime.

Only in passing did the guide mention the historic event most of the passengers most often associated with Newark in general and the Central Ward in particular: the 1967 riots that killed 26, injured 1,100, resulted in 1,600 arrests and left over $10 million in property damage. The fact that the riots had been touched off in good part by plans to construct the very institution that would provide their medical education was mentioned not at all. Indeed, even Udele wasn't aware of the link between her school and the civil unrest that had physically and emotionally annihilated her neighborhood.

Four years after the riots, a twenty-five-year-old newscaster from Ghana emigrated to Newark. Alfred Alexander Ardeyrd Tagoe, the son of a government official in that West African nation, arrived in this country alone, leaving his fiancée, Christophina, twenty, in Ghana until enough money could be raised to bring her over. That would take four years. In the meantime, while he remained in America, they married without her leaving their native soil. In the absence of a groom, Ghanian tradition allows a woman to be married to a proxy chosen by the future husband; Alfred chose his best friend. The proxy stands in for the groom at the wedding, but there, Christophina laughed, his courtesies end. Reunited in Newark in 1975, Alfred and Chris went to a justice of the peace who sanctified their vows in accordance with the marital bylaws established by the State of New Jersey. Soon after, Chris got pregnant and in 1976 gave birth at Martland Hospital to a daughter. The Tagoes named her Udele. When the nurses placed the infant on her mother's stomach for the first time, Chris had a premonition. "This kid is going to be great," she told her husband.

The kid certainly had an early fascination with medicine. "There goes the big hospital," she'd exclaim each time the family car drove past Martland, the first local landmark Udele recognized.

When the erstwhile Seton Hall Medical School, newly minted as the New Jersey Medical College, came to Newark after the riots in 1967, Martland Hospital was, in effect, the school's campus, a happenstance that caused even more suspicion within the already-wary Central Ward. Without a scintilla of evidence, it had long been rumored that Martland's medical staff performed involuntary abortions to reduce the number of black teenage pregnancies.

By the time Udele came to recognize it as the place of her birth, Martland was well on its way to becoming strictly an administrative facility, a transformation completed when University Hospital opened its doors in 1979, two years after NJMS began holding classes in the new medical science building. In every way, the new hospital was architecturally compatible with the grayish-white façade of the medical school, an uninspired, prefabricated cinder-block construction mimicking the design scheme of post–World War II Eastern Europe.

If Udele ever noticed the medical complex in her travels along Bergen Street or South Orange Avenue, she didn't let on. But she continued to fixate on nearly all else that pertained to medicine or the human body. Early on, she was drawn to a friend of the family, a physician who, upon learning the credentials he'd brought to the United States from Ghana didn't meet medical standards, studied furiously until he met the necessary licensing requirements. Recognizing that "Uncle" Christian's profession not only helped people to feel better but also played a role in the birth of her brother, Wesley, in 1983, Udele's response, at the age of six, was to articulate without ambiguity her life's goal: "I want to bring babies out of the stomach."

Although the idea that the daughter of Ghanian immigrants living in one of the poorest sections of a downtrodden city could become a physician seemed, at the time, an impossible dream, Chris and Alfred Tagoe retrospectively understood how it unfolded. The only thing required of Udele was that she remain true to the tenets of a household grounded in principle, hard work and faith.

There was never a time when Alfred didn't hold at least two jobs, and sometimes three. Chris, when she wasn't pregnant or home caring for an infant—there were eventually three, following the birth of Golda Ann in 1985—always worked at least one clerical job while often putting in a few hours a week at another. In the hours when he wasn't driving a limousine or taxicab or repairing cars or doing odd jobs for the post office or telephone company, Alfred found time to study liberal arts at Essex County Community College, a diversion that barely sated an unquenchable hunger for knowledge. Earning money was paramount, the engine Alfred and Chris Tagoe recognized as necessary to elevate their children beyond the environment surrounding them. For there was never any question about where the children would be educated. The Tagoes lived in the Central Ward, but they were not *of* the Central Ward; never, they vowed, would Udele, Wesley and Golda Ann ever cross the threshold of a Newark public school.

Scraping together enough money to purchase their first home in a neighborhood distinguished by poverty and flourishing street corner drug bazaars, the Tagoes managed to carve out a parallel universe. The day the family moved into the two-story attached town house on Muhammad Ali Avenue, Alfred laid down an inviolate rule, informing his children that under no circumstances were they to venture without adult supervision beyond the chain-link fence that surrounded their property. Because Alfred's word was law, the three kids never did; growing up, Udele knew the names of the other residents of the town house complex, but nothing more about them.

"It's not that we think we're better than anybody else," Alfred explained. "It's just that we don't know" the mind-set of many of the neighbors and passersby beyond the fence.

His wife corrected him. "We do know," she said gently. "We do know."

The Tagoes were a unit, with established patterns and rules, setting aside time every weekend for an open family discussion that encouraged everyone to talk about their hopes, fears and the problems encountered growing up in a neighborhood overflowing with despair. A home where all were assigned a specific task led Udele to once complain to Chris that her high school classmates were making derogatory comments about the calluses on her hands. "Tell them you do housework," Chris replied matter-of-factly, reinforcing the parameters that shaped the family.

Coming as they did from an impoverished land, Alfred and Chris often found themselves reflecting on the fascinating and disturbing milieu beyond the fence. They couldn't comprehend how individuals born to a country that offered so much could so blithely squander limitless opportunity. The Tagoes, conversely, fervently pursued the American dream on behalf of their children, preaching the gospel of stubborn perseverance. Udele, in her station as the oldest child, became the first to absorb the lesson.

Displaying an uncanny ability to prioritize, the preternaturally quiet Udele was a quick study, shunning middle school parties or sleepovers in order to complete a homework assignment, preferring, Chris noted, "to be lonely with the books, on a solo mission." The passion to learn came from a father who delivered his sermons on the subject with a healthy dose of realism. "Learning is depressing," he would tell his kids. "It's lonely and it's exhausting. Don't fall asleep; open your eyes and concentrate on what you want. Take nothing for granted. There's no reason to be ignorant about anything. Knowledge is something that no one can ever take away from you. If you have an education, nothing can stop you." While not as vocal as her husband, Chris complemented him by instilling a stubborn resolve. In her every endeavor, she couldn't "stop until I achieve what I want." She made sure her children were the same way.

By the time Udele reached middle school, at St. Mary's Catholic School, the downtown alternative to a public school system so beleaguered it ceded administrative operations to the state, Udele had already absorbed the values instilled by Chris and Alfred. Sensing her potential, the St. Mary's faculty encouraged the seventh grader to take a test designed to identify gifted students who might otherwise wind up mired in a substandard school district. Not long after receiving her test scores, Udele was informed she'd received a four-year high school scholarship from Kent Place, an all-girl private academy in Summit, an upscale suburban enclave fifteen miles and worlds away from the Central Ward.

Every school day for four years she'd rise at 5 A.M., study and then trudge seven blocks to Springfield Avenue to catch the seven o'clock Number 70 New Jersey Transit bus that would transport her from a place of abject hopelessness to one of understated wealth. The days were long: Extracurricular activities like basketball, softball, track practice and dance classes often meant she didn't arrive home until six o'clock. The dividend for the twelve-hour days was an education at a school that guaranteed individualized faculty attention by limiting the total enrollment to two hundred.

While academically suited to Kent Place, from a socioeconomic standpoint, Udele Tagoe could not have been more incompatible with her fellow high school students. Socially, Kent Place catered to the progenies of white suburbia who showed up for classes driving new cars to share classrooms with the offspring of brilliant Asian-Americans reshaping communication technology at the nearby Bell Labs complexes. Udele, one of nine African-Americans at the school, fit nowhere. Neither overt nor vicious, the racism Udele encountered in Summit was subtle, institutionalized and never discussed, except cursorily, at the school's annual nod to diversity, Multi-Cultural Day. Back home, the Kent Place experience only widened the schism between Udele and a neighborhood that would never understand, or appreciate, her ambition.

Still, Udele never forgot—not that Alfred and Chris would have let her, anyway—that she had a gift. Attached to the gift was the inference that, in the future, she might out of obligation return to Central Ward. At an age when most kids are still trying to decide what they want to do when they grow up, Udele Tagoe had already known forever that she wanted to be a physician, a calling that, in her case, carried a burden she didn't pursue but ultimately embraced: role model. "We need doctors, women doctors. African-American doctors," she said resolutely, years later. Before returning home to fulfill her prophecy, Udele needed first to depart. Applying for college, she aimed high— Brown, Johns Hopkins, Emory and Union College—and was elated upon learning she'd been accepted by her top choice, Duke University.

Duke offered two physics courses: one tilted toward premed undergraduates, the other aimed at students matriculating through the engineering curriculum. Not content to accomplish the bare minimum, Udele decided to take both and enrolled in engineering physics during the second semester of her junior year. Up to that point, she'd been more than up to the challenges thrown at her by the estimable North Carolina institution, racking up *A*'s and *B*'s. From the outset, however, Udele recognized that engineering physics had the potential to wreak havoc with her grade point average.

Still, she pushed ahead, motivated by the knowledge that medical schools are filled with engineering students who view the human anatomy as a mere extension of the processes they absorbed as undergraduates. Approaching engineering physics from a different tack, a scientific background, Udele struggled to bridge the gap. Try as she might, she just didn't get it and no amount of tutoring, extra studying or prayer seemed to help.

Although she rarely went anywhere without a cross dangling from her neck, Udele never wore her religion on her sleeve. Growing up, she'd adopted her parents' embrace of the Episcopal Church. In spirituality, as with all other aspects of her life, Udele's was a quiet faith. At Duke, she attended church sporadically. Still, rarely did a day go by that she didn't pray at least once and usually several more times than that. With the final exam in engineering physics bearing down on her, Udele went into full prayer mode: She prayed for guidance; she prayed for intervention.

The morning of the exam, Udele sat down and started going through the questions. Not until page eight did she find one she understood. Udele said another prayer, answered the question and began working her way back through the test. Slowly, the albatross removed itself; the answers started to flow, albeit not enough to alleviate her anxiety.

Adhering to a self-imposed policy adopted to prevent a poor grade in one class from distracting her as she prepared for a test in another, Udele didn't bother to check the physics bulletin board when the test scores were posted the next day. Immediately upon completing her last final, she rushed to the engineering school to get her grade. Standing in the corridor outside the physics department, she was astounded to learn she'd received a *B* in the class, a grade helped enormously by a near perfect performance on the final exam: *A*.

In preparing its premed students for the Medical College Aptitude Test, Duke established the goal of attaining a score of thirty out of forty-five possible points. Udele came very close, netting a twenty-nine. The physics section of the exam posed her the most problems; she could only wonder how different the results, and her life, might have been had she not taken, and persevered, in the extracurricular engineering course.

In high school, Udele had little doubt she'd one day practice medicine in Newark or, at the very least, another urban neighborhood. It took only a couple years at Duke, living and observing the more languid pace of North Carolina, to alter that mind-set. Although remaining in the South meant abandoning her support system back in New Jersey, Udele vowed she'd never return to Newark. When, fortuitously, Wesley began attending a prep school in South Carolina during her junior year, Udele took it as an omen and began to push Alfred and Chris and Golda Ann to join them south of the Mason-Dixon line.

Alfred and Chris were giving the idea serious consideration when something unexpected occurred: Udele's meeting the summer before her senior year with an adviser with the Student Affairs Office at the New Jersey Medical School. The counselor gave her the hard sell, rekindling in Udele the sense of obligation to community, race and gender. The pitch worked so effectively that when the time came to fill out medical school applications, Udele completed and submitted only one. And so the following summer, 1998, as her fellow Duke graduates dispersed to the apartments and careers that signified their entry into the real world, Udele moved back into her childhood bedroom and prepared for the next phase of her life, one that would transpire down the street from where she grew up.

The outside world tends to view medical schools as a citadel through which pass men and women somehow smarter, more insightful and intangibly different from the rest of society. Realistically, medical school enrollment mirrors the general population, attracting the full array of human personalities, from smug to immature, insecure to emotionally undeveloped. As in any setting, the boisterous and cocksure tend to get the most attention. But it is the reserved, like Udele Tagoe, who, over time, cause others to take notice.

One who immediately sensed the potential hidden beneath Udele's unobtrusive demeanor was Dr. Nagaswami Vasan. After observing the Duke graduate during the summer anatomy program, Vasan was delighted to learn Udele had been assigned to his Art of Medicine class when regular classes began in the fall.

Cognizant of the public's perception of physicians as compassion-challenged automatons controlled by managed care bureaucrats, NJMS and other medical schools begin immediately to instill the nuances of proper bedside manner. To drive home the point that the practice of medicine begins on the first day of medical school, many institutions now include an orientation rite once reserved for those who had already earned their medical degrees, the recitation of the Hippocratic Oath. Afterward, as part of a formal ceremony, each student receives the uniform of all physicians, his or her first white lab coat, a symbol of "a doctor's relationship with a patient [that is] is pure and filled with trust." Beyond symbolism, the white coat is a necessity now that medical training almost immediately brings students into contact with real patients in clinics and on hospital rounds.

At NJMS, the first patient, introduced to the students during the first semester in the Art of Medicine, was real only in a composite sense. Splitting the students into groups numbering no more than ten, each week the Art of Medicine gathered participants around a conference table to contemplate, research and solve, step by step, medical crises.

Confronted with hypothetical cases, the future doctors were taught how to think on their feet (although most chose to think with their computers via the myriad health-related Internet Web sites) using problem-based learning (PBL) techniques to diagnose illnesses and formulate a recovery protocol.

Not coincidentally, the first case given to the future physicians at NJMS—childhood leukemia—is arguably the most traumatic news a doctor will ever deliver to a patient and family. Priority number one was to set aside emotion, to turn the diagnosis into a call to arms, a disease to be defeated coldly, efficiently and without remorse. The elimination of the emotional component presents a dilemma as old as the practice of medicine itself. To avoid sacrificing compassion to clinical objectivity, NJMS drills into its charges a touchstone, Unconditional Positive Regard, an aphorism that dictates, above all else, that they listen and empathetically consider the patient's each and every word.

Initially, Art of Medicine illustrated, frighteningly, a naïveté, best exemplified by the discombobulated student unable to locate, on his own body, the general location of the spleen. "Wrong side," Vasan snapped as the student jabbed a finger at the lower right side of his rib cage.

"But . . . but . . . it said the right side," the student stammered ("it" being the Internet Web site he'd consulted the night before), turning the exchange into a "Who's on First" that pitted Vasan and protégé in an elongated dialogue about the difference between the "right side" and the "right" side. The tone was jocular, the subject serious; in the real medical world, innocent mistakes become the stuff of malpractice judgments.

"It's confusing," the student complained. "I don't know if they mean my right side or the patient's right side."

"Always the right side of the patient," Vasan instructed.

Saddled the first semester with biochemistry and cell and tissue biology, the students tended to give problem-based learning short shrift until, one afternoon, a stranger showed up. In real life, she was an administrator in the school's department of education, one of several female volunteers trained by the school to visit the various PBL classes under the guise of being Leslie Simmons, a forty-seven-year-old real estate broker.

According to the prepared script, Simmons had scheduled an appointment for her first routine physical in three years. During the visit, the "doctors" in the room took turns asking questions about her health and lifestyle. When one inquired if she had any STD's, Simmons looked puzzled.

"What's that?" she asked.

"Stay away from the clinical," Vasan reminded.

The students/doctors, using textbooks and the case scenario guidelines distributed each week, debated and then ordered Simmons to undergo a regimen of tests, including a mammography. The following week, a few days before Simmons returned to the conference room, the doctors received word the mammography had revealed a spot in the left breast.

"OK," said Vasan when the class reconvened. "Do we give her the good news first or the bad news first?"

"The bad news first; we don't want her to leave on a negative note," said Shawn London, summoning a strategy he'd read in a textbook.

"There are a lot of things you don't learn from books," Vasan sighed, prompting a high-spirited twenty-minute debate that ended with everyone agreeing that, in this province, there were no right or wrong answers. Dividing the workload among themselves, the class prepared to deliver to Leslie Simmons the results of her physical. The task of telling the patient about the mammography results was given to Udele Tagoe.

Opting for the good news/bad news approach, the first student/doctor told Simmons to stop smoking, the second suggested a diet to maintain a relatively low level of cholesterol, the third recommended an osteoporosis screen. The tension built, broken only by Simmons, who asked how to counter the inevitable weight gain should she quit smoking and whether her insurance would cover the osteoporosis screen ("A common question and one that these days you'd better be prepared to answer," Vasan pointed out). Silence descended; all eyes turned to Udele.

Nervously, Udele glanced at her notes and then cleared her throat. Making small talk, she emphasized to Simmons the importance of breast self-examinations and a regularly scheduled mammography, a stall for time until the segue into what needed to be said.

"And how was your mammography?" Udele asked.

"Freezing."

Udele drew a deep breath, her mouth went dry and she reminded herself that the woman seated across the table was no more a real patient than she was a real doctor. The admonition did little good; this felt real, very real. I hope she doesn't freak out on me, Udele thought in the instant before she cleared her throat and began: "Your mammography showed a shadow in the left breast that is larger than it should be."

Simmons leaned forward, alarmed: "What does that mean?"

"Uh, well, it could just be a shadow or—"

"I want to know, what's wrong? Do I have cancer?"

"We won't know what's wrong until you have a needle biopsy."

Simmons's tone assumed a new urgency. "You must have some idea. What else can it be if it's not cancer?"

To her surprise, Udele was suddenly centered by a surreal calm. Looking straight at Simmons, she said, "It could be a cyst or a malignant growth, or like I said, it might just be a shadow on the picture, sometimes the angle—"

"There must be something to biopsy if you want me to make sure it's not cancer. This is the same thing that happened to my aunt and it killed her. . . ."

"I just want you to understand that just because there is a shadow it does not mean we found cancer there. The important thing is that you don't get ahead of yourself," Udele said firmly, enjoying the sensation of power shifting from patient to physician. It no longer mattered if Leslie Simmons challenged, cajoled or teetered on the brink of hysteria because, ultimately, Udele Tagoe was in control.

She'd changed that afternoon, no doubt about it, a metamorphosis from student intimidated by circumstance to woman of stature, a caring physician capable of easing Leslie Simmons's apprehension with the reassurance that, should the opportunity present itself, her every medical need would be meticulously addressed. In the weeks that followed, Udele replayed the exchange over and over, looking for perspective, never losing sight of the fact that what passed as pure theater in the Art of Medicine would, in the not-too-distant future, be the antithesis of academic exercise.

CHAPTER 7

———————•———————

T he first order of business when Vasan's Art of Medicine class reported to the conference room seven days later was to review biopsy results that revealed the shadow inside the fictional Leslie Simmons's left breast to be a benign cyst. The administrator playing the role of patient did not appear personally to receive the good news, thereby denying Udele the honor of delivering it or the opportunity to reiterate the importance, given the history of breast cancer in Simmons's family, of an annual mammogram coupled with regular self-examination.

Redoubling the effort to convince her to quit smoking while stress-ing the need for a healthier diet, the students gave the patient Uncon-ditional Positive Regard in a minimum amount of time, since the new week brought another theoretical medical crisis visited upon yet another hypothetical patient.

While the crux of the Art of Medicine was to prepare first-year stu-dents for a career that requires its practitioners to perpetually adjust as they grapple with the unique personalities and health complaints lurking behind every examination or hospital room door, the course also served as a sound foundation for what lay ahead in the second semester. The Class of 2002 learned just how ephemeral gross anatomy could be when, four days after triumphantly removing the lungs, they in essence started from scratch.

"Is our guy still here?" Sherry asked, unhinging the cabinet doors Monday afternoon. Had he not been, no one would have been more surprised than Udele, who had seen him the night before when she'd come in to brush up on the lungs and thoracic cavity.

Sherry, anxious to further investigate the tantalizing surgical clues uncovered the week before, had all but forgotten about the lungs.

Were it not for Vasan's stern advisory to the entire lab—"Don't remove the heart right away. Identify the structures around the heart first. Take your time, forty-five minutes or so. I really want you to understand the clinical anatomy of the heart"—she would have been perfectly content to immediately cut away the arteries, veins and other connective tissue holding the heart in place.

Udele shouldered the brunt of the scalpel work, revealing the pericardium, the sac enclosing the heart, stopping periodically so her lab partners could track her progress. Every cut exposed more sutures, reinforcing the earlier evidence that the donor had suffered major arterial blockages.

"That's what you get when you eat at Burger King and McDonald's," Ivan said, getting no argument.

"This is the pericardium?" Jen asked. It looked nothing like the depiction in the books. She placed two gloved fingers under the sac and probed, feeling for the arteries and veins the *Grant's Dissector* pledged would be there. Withdrawing her hand, she held it aloft admiringly, observing, "You know, this is the best tool; it's smooth, it's not sharp and it can feel everything perfectly."

Compared with the precision necessary to preserve the circulatory system surrounding the heart, the first week's dissection had been haphazard, crude, a mad rush through fat and extraneous tissue. Then came the unit's second phase, an exercise that demanded a higher level of exactitude and intense concentration that forever changed the dynamic. "This makes me really nervous; you cut away the wrong thing and your specimen is shot," Sherry admitted, her eyes moving back and forth from the thoracic cavity to the *Dissector* perched on the bookstand attached to the end of the table, parroting a musician dividing attention between conductor and score.

Watching Sherry struggle, Jen briefed the table about her weekend visit to a group of law school friends in Boston and the consequence of disclosing to them details about the glob of fat stuck in her hair the first day of lab. "After that, they acted like they didn't want to hang out. I guess they were afraid I'd have some extraneous body tissue stuck to me," she said, vowing in the future to limit lab-related discussions to fellow medical students. The flip side of the intellectual development fostered by gross anatomy, Jen lamented, was the penalty exacted on her love life. "It's not a good time to be single; I guess there won't be a lot of dates until this class is finished," she predicted.

As Udele trimmed the last vestiges of tissue from the pericardium sac, Jen peered over her shoulder. "Oh, it's a pretty heart. A pretty dissection," she enthused.

But it wasn't a pretty heart, it was an ugly heart, mottled by disease, scarred by surgery, obscured to the point that crucial components Table 26 needed to analyze were practically unidentifiable. Jen raised her hand and looked imploringly at Zolton Spolarics.

"You guys are always asking the most difficult questions," the instructor said good-naturedly. Given what he'd seen previously, Spolarics wasn't surprised at what Table 26 had uncovered. "You can't expect normal anatomical conditions here," he noted. "You'll see the ventrical walls, and the arteries, but I don't expect you'll see much more."

"In other words, we have a dud," said Sherry.

Spolarics instructed them to leave the heart in the pericardium and directed Sherry, Udele, Ivan and Jen to Table 24, where Cary Idler, ably assisted by Leslie Pooser, had removed a heart so perfectly formed that Frank H. Netter could have used it as a model for his *Atlas*. Sherry sighed. "It makes you appreciate what a mess our heart is when you see a good one."

"Theirs is so good, ours is so horrible," Udele added in sad concurrence.

In the grand tradition that began with rounds to examine tumor-ridden lungs, Table 26 became a magnet for classmates beguiled by deviation. Leslie Pooser took one look and grimaced: "I'll tell you what, I'm not smoking and from now on I'm not drinking, um, very much. And I'm not going to Burger King anymore. Did you see between those ribs? The intercostal? Barbecue sauce. I'm bringing my mother in here, she needs to see what that stuff can do to you."

While her classmates marveled at the lab's disparate hearts, Jennifer Heimall—at Table 28—experienced an epiphany of a different sort. Emotionally braced for the task ahead, she'd remained stoic and businesslike during the initial cut and removal of the lungs. For Jennifer Heimall, heavily involved in student government and seen as one of the leaders in the Class of 2002, maintaining appearances was important. The policy of studied detachment implemented so efficiently the first week began to crumble when Jennifer's table severed the aortic arch, the pulmonary trunk and the inferior vena cava. As a lab partner reverentially lifted the heart from the pericardium, Jennifer's emotions cascaded upward from inside her own chest. Swallowing hard to get rid of the lump in her throat, she considered the heart, glistening on the table. It was, she thought, absolute in its beauty, perfect, "like a bonsai tree" where the pulmonary veins and arteries branched out.

"Oh, my God." The phrase, exclamatory, escaped involuntarily. A lab partner looked imploringly at Jennifer, worried she was about to vomit.

"You OK?" he asked. Jennifer was fine; more than fine, really, as, for the first time, the enormity of what she was doing, what they all were doing, crystallized. The cadaver, with towels covering its every part save a chest now emptied of its heart, was a human being who'd lived a full and, she assumed, productive life.

"You OK?" the lab partner repeated. Jennifer nodded, smiled and picked up the syllabus to see what needed to be done next.

Jennifer Heimall was far from the first medical practitioner to become flummoxed by the miraculous power of the human heart. The earliest physicians, in fact, were so infatuated with the organ that it set back the understanding of medicine by centuries. Believing the "seat of intelligence" resided in the heart, the physicians who followed in the wake of the Father of Medicine, Hippocrates (ca. 460–ca. 377 B.C.), saw no need to further investigate the anatomy by dissecting the human body. That philosophy of benign neglect, subsequently propagated by religious doctrine equating dissection with desecration, generally carried the day until early in the second century A.D. when Galen of Pergamum composed the sixteen treatises, most famously *On the Uses of the Parts of the Body of Man*, that earned him the sobriquet, in the manner of Hippocrates, Prince of Physicians.

Circumventing religious prohibitions regarding human dissection by working with animals, Galen identified the chambers of the heart and correctly attributed the circulation of blood to a system of arteries and veins. Thoroughly rejecting both Judaism and Christianity, the physician nonetheless ascribed the ultimate source of the vascular system to a "spirit," hypothesizing that the deity could well reside in the liver.

The passing of Galen marked also the death of cogitation that might have advanced modern medicine beyond his embryonic theories. Not until the Middle Ages, when the papacy and Italian government sanctioned postmortems at the medical schools in Bologna and Padua, did dissection emerge as the primary means of understanding the body. Nearly two millennia after Hippocrates lived, Dr. John Siegel, professor and chairman of the Department of Anatomy, Cell Biology and Injury Sciences at the New Jersey Medical School, riveted the first-year students during his opening lecture with a chronicle of anatomy and dissection. And there ended the Class of 2002's intersection with medical history.

What Galen and Hippocrates had done on behalf of their education was of no interest to Table 26, which was far more troubled in the second week of 1999 by its status as the final table in Lab C to remove the heart, a situation Jen Hannum blamed on her inability to pinpoint the recommended aortic landmarks. "I'm sorry to keep going over this, but I'm just not getting it," she apologized.

Sympathetically, Sherry demonstrated how multiple bypasses had altered the anatomy, redirecting the logical flow of blood so that the veins and arteries looked nothing like the textbook illustrations. Vasan reinforced that appraisal, using a probe to show how surgeons had routed the gastric artery through the diaphragm and into the heart.

As usual, the appearance by Vasan drew an audience of awestruck students unintentionally reenacting the tableau created by Rembrandt in his famous conception *The Anatomy Lesson of Dr. Nicholaes Tulp.* Completing the presentation, Vasan downplayed the adulation, quipping that he'd derived his knowledge of bypass techniques from The Learning Channel. "They made a mistake describing the anatomy during the show," he hastened to add.

The assemblage dispersed and Vasan went back to work, pinching the connective tissue and arteries while Jen, fearful that she might make a mistake in front of the revered course coordinator, deftly sliced through the tissue, unburdening the heart.

"Before we take it out, can we go over the vessels one more time so

I don't have a heart attack?" Sherry asked. The rest of the table stepped back and allowed Sherry to sketch a quick diagram. Already far behind the progress made by the rest of the lab—some of the tables had removed their hearts over an hour before—Table 26 didn't begrudge Sherry a few extra minutes.

When, at last, Udele slowly began lifting the heart from the pericardium, it snagged. "There is some sort of something here that won't tear," she said, frustrated.

"I feel like we're delivering a child or something," Jen volunteered, drawing a laugh from Sherry, able to testify that childbirth was often more expedient.

"OK, last house call for your table," said Vasan, making a precipitous arrival.

Vasan studied the problem and frowned. "This is the ascending aorta and this . . ." The course coordinator hesitated. "Well, it's hard to say what this is. When [the surgeons] do something like that, you lose some of your anatomy." Again, Vasan clamped and Jen cut. "It's like doing surgery all over again," he said, reaching in, plucking the heart from the sac and placing it on the table.

"All right!" Ivan exclaimed, an enthusiasm that diminished as soon as he and the others took measure of what Vasan had removed. Outside the thoracic cavity, the heart looked even worse than it had inside the chest. Vasan made a feeble attempt to exhibit the ventricles before stopping abruptly to summon an alternative.

"Cary, can they borrow your heart?" he called out to Cary Idler. Seconds later, the heart from Table 24 sat in bold contrast next to the heart from Table 26.

"Oh, that's a beautiful thing," Sherry marveled, tracing with a finger the left descending artery. "Damn, what a beautiful dissection."

"And this is the great [cardiac] vein. Pull on it. See? It moves over here," said Jen.

"I'm impressed," Ivan countered.

Jen looked abashed. "Don't be. Someone showed me. You think I came up with this on my own? Don't give me so much credit."

Five months into medical school, Jen still sometimes felt like an interloper, an outsider who'd stumbled into someone else's party and decided to stay. Each afternoon she stood elbow-to-elbow with three people who couldn't remember a moment in their life when they hadn't aspired to becoming physicians, as if medicine had divinely plucked Udele, Sherry and Ivan from preadolescence and decreed that one day they would be doctors. For Jen, every day brought another reminder that medicine had not chosen her, she had sought out medicine.

As a kid growing up in Little Egg Harbor Township, just north of Atlantic City, athletics—and not a future professional career—had distinguished Jen from the rest of the preadolescent horde. Even in middle school, she was a flat-out jock, competing, sometimes simultaneously, on three different basketball teams, including one squad that regularly trounced the varsity high school team in scrimmages.

It was a harbinger of what was to come at the high school level, where Jen's Little Egg Harbor Township Lady Eagles won two state championships and barely missed out on a third. With Jen playing point guard the team went 61–10 in four years and was ranked at one juncture the fifth-best high school women's team in the country. In her junior year, Jen was named the most valuable player in the conference, an honor that usually went to someone with a high scoring average. Not limited to basketball, her athletic abilities extended also to a fall schedule of field hockey; in the spring, she played catcher on the softball team—Jen's best sport, in the opinion of her father, Jay.

Joanne Hannum, the family archivist, meticulously maintained the scrapbook that tracked her daughter's accomplishments. In news photographs, Jen is the Lady Eagle easiest to identify. Whether the rest of the team exults in victory or anguishes in defeat, in each photo Jen is the picture of impassivity, her face betraying nothing. Happy or sad, she kept it all inside whether on the basketball court or at home. The only time Joanne ever saw her daughter weep was when a calculus test forced her to miss a basketball game; her father couldn't recall a moment, outside of childhood, when his daughter cried.

The single-minded determination and fierce competitiveness first became evident in the fifth grade when Jen, having just received a *B* on a math test, returned home livid, swearing it would never happen again. It never did: She graduated from high school with a cumulative 4.41, a grade point average bolstered nearly half a point by excellence in honors classes. At the senior honors ceremony Jen received so many academic and athletic awards, twenty-seven in all, that a kid in the front row finally offered to give up his seat so she could more easily access the stage each time her name was called.

Four years before that ceremony, Joanne Hannum had a foreshadowing of what lay ahead. Jen was in the eighth grade and had just brought home another perfect report card; Joanne turned to her husband and said, "You know, I don't have a clue or anything about college, but I think I better start finding out."

College had never been part of the equation in a home where the oldest of the two Hannum children excelled in vocational arts and decided halfway through high school to enlist in the U.S. Marine Corps. Nor had Jay or Joanne seriously considered the possibility of pursuing a secondary education. Married young—Joanne was sixteen, Jay only two years older—Jay became a fireman while Joanne worked as a legal clerk. Then came Jen, who, by the time she hit high school, knew not only that she'd be going to college but also what her major would be.

Acknowledging Jen's lifelong affinity for sports and science, Joanne thought her daughter should pursue a career as a physical education teacher or, given her love of animals, veterinarian. Jay ruled out the latter, recalling how Jen bolted from the room at the first sight of blood whenever he tuned to the surgeries broadcast on The Learning Channel. One night, Jay brought home a book detailing career alternatives and, together, father and daughter explored the options. "Why don't you become a pharmacist? They make lots of money," Jay suggested, tapping into Jen's fascination with chemistry. The allure of wealth didn't clinch it, the prospect of a life immersed in chemical compounds did: Jennifer Hannum, all of fourteen, decided on the spot she would be a pharmacist.

By her senior year in high school, Jen felt the tug between the two things she enjoyed the most: school and sports. While four of the Lady Eagles' starters accepted college basketball scholarships (one went on to play on the NCAA championship team at the University of Connecticut), Jen held back after receiving an offer, from a school of pharmacy no less.

The day she sorrowed over missing the basketball game in order to take the calculus test, Jen had been given the option of taking the exam at a later date, an alternative rejected on two counts: Not only was it unfair to the rest of the class, which had to take the test on schedule, but it also violated a personal axiom that academics always come first.

To accept a scholarship at the college level, Jen feared, would elevate her debt to basketball while diminishing her commitment to school. Still, the tender from the Philadelphia College of Pharmacy enticed with its ability to ease Jay and Joanne's financial burden. Jen talked it over with Joanne, who agreed that the practice and travel necessary to compete at the college level would detract from her daughter's education. The talk was more reinforcing than persuasive; by then Jen had already made up her mind: The time for playing games had passed and the moment for the more serious pursuit had arrived. The next day, she submitted her application to Rutgers University in New Brunswick.

Jen took immediately to the school designated New Jersey's state university. Best of all, her GPA was none the worse for a lifestyle that now included the requisite collegiate partying. Between her freshman and sophomore years, she returned to Little Egg Harbor and a summer job at the local hospital, where, curiously, Jen felt pulled toward a profession other than pharmacy.

That summer, it was Jen, and not her father, who made the real-life medical programs on The Learning Channel part of the family's television regimen. One evening, the show followed a neurosurgeon seeking the cause of a speech impediment by asking the patient, conscious during the open cranial surgery, to speak each time an electrode was attached to a lobe of his brain. Jen was transfixed by the excruciating exercise in trial and error. Finally, the neurosurgeon pinpointed a lobe that rendered the patient speechless. "That's it," said the physician, ordering the patient anesthetized for the removal of the lobe.

The instant the surgeon's probe struck the problematic lobe, Jen's "jaw hit the ground." Afterward, she remained paralyzed with awe. For five years, she'd wanted nothing more than to be a pharmacist. Now, for the first time, she began to seriously gravitate in a different direction.

Keeping her own counsel, Jen vacillated, torn between allegiance to a goal established when she was fourteen years old and the gradual realization that, ultimately, being a pharmacist might prove to be an insufficient challenge. During her senior year at Rutgers, the internal debate became so combustible that Jen broke the self-imposed embargo and floated the idea past some other pharm majors. Many of them, it turned out, were contemplating the same move. Hearing this, Jen shoved the idea aside, convinced the whole thing was a phase.

At the end of her fourth year—pharmacy majors spend a fifth year doing rotations in hospitals and clinics before receiving a degree—Jen concluded the phase wasn't going away. That spring she took the MCATs and then departed for a summer research fellowship at Michigan State University, as far away from New Jersey as she'd ever been.

When Jay and Joanne journeyed to East Lansing for a visit, Jen sat them down for a heart-to-heart, informing them she'd taken the MCATs, had done well (thirty-one points) and, upon her return to New Jersey, would begin the process of applying to medical schools.

Medical school?

No sooner had the words been spoken than Jay and Joanne flashed back to the kid who fled the room whenever her father flipped on The Learning Channel. Once the shock began to wear off, Jay began dealing with the practicalities: Getting Jen through five years of undergraduate study had been difficult enough; four years of medical school—not to mention a minimum three-year residency and internship—seemed financially impossible. Jen assured them she would take care of it and they promised to help as best they could.

While his daughter certainly had the brains to get through med school, Jay still wasn't convinced she had the stomach for it. A few weeks later, a PBS documentary on the rigors of medical school gave Jay further pause. Despite providing only the slightest glimpse of what occurred during a mandatory course for all first-year med students, the program's visual information was sufficient to cause Jennifer Hannum's father to express concern that his daughter might not survive the rigors presented by gross anatomy.

Jen's first exposure to medical school proved her father prescient in one regard: The application fees alone exceeded $1,200. George Washington University in Washington, D.C., was her first choice, and it remained atop the list until Jen received the informational packet detailing four-year tuition costs that exceeded $200,000. Unless a scholarship came through, she resigned herself to being an in-state student at one of the two medical schools operating under the auspices of the University of Medicine and Dentistry New Jersey. Since she'd already lived in New Brunswick, Jen applied to UMDNJ-Robert Wood Johnson Medical School with ambivalence. That left one alternative, the UMDNJ-New Jersey Medical School in—and Jen couldn't believe she'd ever live there—Newark.

Dr. George F. Heinrich, the assistant dean for admissions at NJMS, was very familiar with Newark's reputation among prospective students. Overwhelmingly, those who reject the school's offer of admission name the city of Newark as the reason. Heinrich typically countered the merest hint of indecisiveness by citing the diversity of clinical training inherent in an urban environment. Most of the time the imploration was wasted on deaf ears.

Each year, the New Jersey Medical School receives 3,000 applications for admission. Of those 3,000 applicants, 750 are brought in for a personal interview or audited by a seventeen-member admissions committee. Of the 750 finalists, approximately 180 are accepted. On average, NJMS students score a 30.2 on the MCAT. Nationally, an average of 38,000 applications are submitted for 16,000 medical school admissions and the mean MCAT score is 29.8.

Jen's initial visit to NJMS was marred by the interviewer, who chastised her for not stipulating that the pharmacology research fellowship had occurred at Michigan State and not the University of Michigan. "What do I know? I'm not from Michigan. She asked where I'd done my research, I said, 'Michigan.' I'd meant I'd studied in the State of Michigan, that's all." (Sherry had been equally disturbed by an interviewer of a certain gender and age who wondered, should she be accepted for admission, if the wife and mother would be able to "take care of her husband." Sherry replied in the affirmative, a response that triggered the rejoinder, "Medicine is a jealous mistress.")

Focusing on motivation, the hour-long interview is a key component in the decision about who will be admitted to NJMS, said Heinrich, but he stressed it is not a make-or-break proposition. One person's opinion, he pointed out, "can't destroy someone, nor can it get them in."

When all the variables—grade point average, MCAT score, essay and personal interview—add up to an admission, NJMS diverts from the tradition of written notification, deferring instead to Heinrich, who takes great pride in delivering the good tidings personally.

The telephone rang just as Jen, Jay and Joanne were preparing to leave for a holiday party. George Heinrich was on the line. "Congratulations!" he boomed when Jen picked up the phone. The moment was ethereal, it was Christmas Eve, her father was dressed as Santa Claus, they were halfway out the door and, now, a stranger was on the telephone informing Jen that she'd been accepted to medical school.

Jen received the news with all the excitement of someone who'd just been served soggy cereal for breakfast. "It's Christmas Eve, doesn't he have anything better to do?" she said to her parents after Heinrich hung up. As they walked out the door, Jen made Jay and Joanne promise they wouldn't breathe a word of the telephone call to anyone at the party.

Having switched her life's direction at the midnight hour of her undergraduate education, Jen nonetheless knew immediately the niche of medicine in which she most wanted to practice: In all of medicine, there was but one place that could match the adrenaline rush of athletics: the emergency room, where every hour of every day brought with it a new challenge. Jen wanted to be an ER physician. And in the pressurized cauldron encountered each afternoon in the gross anatomy lab, Jen sensed right away she'd arrived at ground zero, basic training for what lay ahead.

If the afternoons Jen spent in lab were enervating, the mornings spent listening to an instructor drone on about physiology were the exact opposite. After two hours of physiology, no one was much in the mood to hang around another sixty minutes for the gross anatomy lecture. By the semester's second week the GA lectures had already become superfluous, attracting no more than 75 percent of the class on a good day.

In addition to following soporific physiology lectures, the GA faculty also had the disadvantage of delivering their remarks at noon, when blood sugar tended to hit the daily low. The situation, ironically physiological, dovetailed into the second problem, a cavernous and dark lecture hall that lent itself to sleep. And nodding off was precisely what a good percentage of the students did.

No one slept when Siegel lectured and, given his reputation for erudition, few cut class. As a result, the lecture hall was filled to near capacity when, in the middle of the heart dissection, the chairman of the department appeared at the lectern to put into perspective everything currently transpiring in the lab.

Drawing on his stature as a renowned trauma specialist, Siegel told of the day he'd received on his pager a message: "Patient shot with arrow." In parenthesis, the messenger had added, "Yes, an arrow." Siegel hurried from his office in the Medical Science Building to the emergency room at University Hospital. He was not alone; as word of the injury spread through the hospital, at least forty other doctors and nurses wandered by to see for themselves because, "as you know, we don't have a lot of Indian attacks in Newark."

The patient had wounded himself in a suicide attempt. Siegel couldn't explain why he had used a crossbow, not the most efficient method to kill oneself. But the chairman was able to detail what happened once the arrow pierced the cardiac chamber: "The pericardium sac surrounding the heart fills with the blood that leaks out from the heart itself, limiting the ability of the heart to dilate in order to accept blood from the peripheral vessels. The patient goes into shock when blood can't enter the heart, the blood pressure drops . . ." Siegel explained, illustrating his dissertation with slides taken while he removed the arrow from the patient's heart.

In the audience, not an eye was shut in slumber as, before them, Siegel provided an image of what they were now witnessing in lab, a reminder that their undertaking was more than a grand experiment or a quest for a grade. Midway through the lecture came another reminder: the muffled thud of rotors as the New Jersey State Police MedEvac helicopter landed on the roof pad of the adjacent hospital with a patient bound for the trauma unit following an automobile accident or an equally horrific event.

Siegel concluded by noting the sutures that closed the hole in the victim's heart soon brought full recovery. "And he did just fine," said the chairman. He paused for effect before adding, "Except psychiatri-cally, of course."

The chairman's appearance in the lecture hall had the impact of a halftime pep talk; an hour later, the class fairly bounded into the lab to complete the dissection of the chambers and orifices of the heart. Before Table 26 could start, they first needed to clean their heart by the most ordinary means—under a faucet.

"Just like washing out a chicken or a turkey," said Sherry.

"Thanks for the thought," Ivan retorted. "I think I'll skip Thanks-giving this year."

Jen placed the heart, glistening with water, on the table and handed the scalpel to Udele. "I trust you more," she informed her lab partner.

"Yeah, we haven't done this yet. You have," Sherry added.

"Yeah, we'll do it when we get down to the abdomen," Jen said.

"That's good, because I won't know what I'm doing when we get down there." Udele grinned, using the scalpel to adroitly prove Jen right.

Udele's sureness unnerved Jen and Sherry, again frustrated in the attempt to locate the orientation points outlined in *Grant's Dissector*. The ventricles, the atriums, the aortic valvulae—all of it looked the same.

"I'm sorry, we're going to have to go slower," said Jen, suggesting that they "identify one part at a time. Then, when we've all looked at it, we'll move on to the next one."

As long as the others allowed her to maintain control of the scalpel, Udele didn't mind the phlegmatic momentum. Methodically, she tugged at the heart tissue with the forceps, stopping every thirty sec-onds to provide Sherry, Jen and Ivan the opportunity to recite the litany prescribed by the *Dissector*.

When Udele indulged a habit of sometimes getting ahead of her lab partners, one in particular lapsed into third person to call her on it: "Wait a second, you're moving too fast for Jen!"

Recitation being the soul of memorization, Sherry and Jen main-tained a steady dialogue, an audible tour for the benefit of themselves as Udele disassembled the vena cava . . . inferior vena cava . . . right semilunar cusp . . . left anterior papillary muscle . . . interventricular septum. . . .

"If I don't have this stuff shoved in my face, I'll pretty much forget about it. I know, it's pretty kindergartnerish, but that's the way it is," said Jen, without apology.

Socially, the lab settled into a pattern. Having already concluded they disliked one another, the lab partners at some tables went drone-like about their business, wasting not a syllable on extraneous discourse. Other tables enjoyed each other's company too much and consequently spent the afternoon cutting up in a manner not prescribed by the anatomy department. At Table 26, an easy friendship started to emerge, especially among the three women.

"Who wants to be a surgeon?" Jen asked, apropos of nothing.

"Maybe OB or pediatrics," said Udele, her eyes still riveted to the heart.

"ER," said Jen

"Too hectic," Udele countered.

"I like that, hectic. It's not always like that, though. The car accidents and putting people back together, yeah, but the rest is just people puking and pissing."

"I'd like to work with kids," Udele ventured.

"The problem with peds [pediatrics] is that you have to deal with the parents," Sherry noted.

"Deliveries, too," said Udele.

"You ever spend any time in an OB ward? Women whining, women complaining, constantly. I know, I was one of them," said Sherry.

"Look at Udele," Jen exclaimed, returning to the business of the heart. "She's a human TPA [tissue plasminogen activator] busting up those clots."

Sidling up to observe her lab partner more closely, Jen innocuously nestled into the crook at the elbow of the cadaver's outstretched arm. "And you said you couldn't get a date this semester," Ivan remarked.

"He's the only one who doesn't think I stink," Jen said, accepting the cadaver's embrace.

Time had vanished into a vacuum, warped by a voluminous amount of information. The previous week there'd been the twenty-eight different parts of the lungs, not counting the myriad supporting structures to learn. Now a prolific accumulation of ventricles, veins, arteries and nerves necessary to maintain the percussion of the human heart had been added to the catalog. And still they were only halfway through the first unit of the class. By the first exam, they would also be expected to know what lay inside the arms and the back.

It was difficult to believe that barely a week before only a handful of the class had ever had the experience of cutting into a human body. Christmas, celebrated just three weeks earlier, seemed as distant as the Fourth of July.

The first two weeks of lab had been frenetic, fast-paced, an ever-changing menu of surprise and intrigue. Not unlike, Jen imagined, what transpired day and night in an emergency room.

CHAPTER 8

———————————•———————————

Going into gross anatomy, those who fretted about such things scanned the course syllabus, taking careful note of the linchpins that threatened to inflict damage to the psyche in one form or another. Topping the list, naturally, was the first cut, followed, in order, by the first exam, the dissection of the face, the second exam, the dissection of the genitalia and the final exam. In the event that lineup didn't do the trick, the Department of Physiology more than happily obliged the general malaise by scheduling an exam every other week for the duration of the semester, for a total of six.

In calculating the potential for psychological injury, few students took into account an event as traumatic, in its own way, as the initial cut: the maneuver necessary for dissection to shift from the thoracic cavity to the muscles of the back. One who had given the matter thought, at least subconsciously, was Leslie Pooser.

On the night before each table turned its body from the supine to the prone position, Leslie dreamed of the procedure. The precise details of the dream evaporated when she awoke, but the sense of dread that something bad was about to happen lingered. "I don't want to be flipping nobody over," Leslie declared upon entering lab the next afternoon, articulating the unspoken fear of countless others.

During the first two weeks, Leslie's table, number 24, had been careful not to disturb the towel covering the cadaver's face. They weren't alone. Even Table 26, where the lab partners professed to have no qualms about such things, dared not remove the face towel after Jen and Udele's first-day glimpse of the man who had donated his body.

Unfortunately for those students who did have qualms, the dissection could not progress any further without exposing the entire body, effectively shattering the illusion that they were working with a body designed and manufactured solely for the purpose of expediting their education. With the towels removed, the body could no longer be viewed as a disparate entanglement of bone, muscle, nerves, arteries, veins and vital organs. Seeing the cadavers in toto, the students were confronted with human beings who appeared as they did when once they walked among them. Beyond that, the unveiling forced the students to look at the face of their cadaver, an act many had steadfastly avoided.

Working through her fear, Leslie Pooser took the liberty of removing the towel from the face of the cadaver at Table 24, an act that left her rather surprised. In every corner of the lab, similar dramas unfolded: Table 30 discovered a tangible reminder of their cadaver's humanity, the bobby pins she'd evidently fastened to her hair in the hours before death; Annette Pham at Table 23 "freaked out" when she realized the man whose chest she'd just spent two weeks dissecting had been someone's grandfather.

Christine Ortiz's table was no different from all the others: At the conclusion of lab each night, she and her lab partners ceremoniously repacked the chest, reinserting the heart, the lungs, securing the vital organs with the breastplate. Using the outermost flaps of skin to cover the contents, it was almost as if they were reconstructing, on a grand scale, a Visible Woman, the plastic model that has introduced anatomy to thousands of children.

At Table 42, the rite took on even more significance after one of the partners, on the first day of lab, deemed the cadaver a "piece of shit." The blatant insensitivity rallied the rest of the team into a unit that banded together to protect their own, and the donor's, best interests. Sensing hostilities yet to come, Christine, Jen's roommate, immediately aligned herself with the only other woman at the table, Leah Schreiber, a student who had taken a most circuitous route into med school.

A troubled teenager, Leah spent high school shuttling between the homes of her divorced parents, running away from both. After receiving a hard-won diploma, Leah went to work at the local shopping mall, eventually working her way up to assistant manager of a clothing store, an accomplishment she likened to hitting "rock bottom." Reexamining her priorities, Leah abandoned the mall and took the first steps—although she didn't know it at the time—toward medicine.

Setting out to become a lab technician, she'd no sooner completed the requisite courses at the local community college than word came down that most labs were demanding four-year degrees. By then it didn't matter: A part-time lab job had already convinced Leah that blood-typing was only slightly more challenging than selling clothes in the mall. Since "some people hate hospitals, but I've always been attracted to them," Leah enrolled in the nursing program at Rutgers.

There Leah met the boyfriend, a premed major, who persuaded her that, intellectually, she was better suited to becoming a physician. Leah didn't argue; even while working as a lab tech at a hospital near Atlantic City, she'd been so fascinated by the human body that she passed up a dear friend's wedding in order to accept an invitation to observe an autopsy. Urged by her boyfriend, Leah began applying to medical schools. Seven long years after she quit the mall, taking off the occasional semester to finance her education, Leah Schreiber received the call from George Heinrich.

Even then, "I was very apprehensive that I made the wrong decision. I mean, I'd already made every wrong decision when it came to becoming a doctor: I almost dropped out of high school, I worked in a mall, I wasn't serious in college. I'd made every turn to lead me as far away from this as possible."

Understanding that her presence in medical school represented a personal miracle, Leah's appreciation for the gift bestowed by the stranger lying atop Table 42—a "pleasant-looking" woman, "you could tell she was someone's grandmother, you could see her knitting, taking care of the grandchildren"—was transcendent. To Leah, the cadaver was never just an object to dissect but always, especially in the wake of the sentiment expressed by her lab partner, a person in need of care.

Every night, she took it upon herself to perform the ritualistic repackaging of the thoracic cavity. In a body ravaged by cancer, Leah and Christine had discovered a heart with a near-perfect aorta. Acting to share the perfection with future classes, an impressed Vasan absconded with the heart, exchanging it for two other hearts secured from the supply room freezer. One, two, it made no difference to Leah: Without irony, from that point forward she placed two hearts into the chest at the conclusion of lab. All was well until the flip, an event that permanently altered the established sense of order at Table 42, when the lovingly packaged contents of the thoracic cavity spilled onto the table and out over the floor with a resounding thud.

Reflexively, Christine gasped, covered her mouth and sprinted out of the lab with Leah in close pursuit. In the corridor, they silently embraced, unable to articulate a mutual feeling of lost innocence, knowing the difficulty of ever again looking at the cadaver in the context of her life. Reentering the lab, Leah picked the hearts and the lungs up off the floor and, without a word, placed the organs in the translucent pink plastic refuse bags set aside for that purpose.

Simultaneous to Table 42 grappling emotionally with the flip, Table 26 encountered a more technical problem brought on by an injury to the person key to their lifting their cadaver. During an ATP/2002 basketball game the previous week, something had given out in Ivan's knee, sending him sprawling onto the NJMS gymnasium floor. With Jen's assistance, he limped next door to the University Hospital emergency room to begin an endless wait that became Jen's first brush with the "real Newark."

Being a medical student, especially a first-year who rarely had cause to leave the building, made it easy for Jen to forget her surroundings. Newark was a place Jen experienced only in passing, her exposure to storefront check-cashing emporiums, burned-out buildings and young men aimlessly lingering on street corners limited to the view from her pickup truck on the way to school.

After parking the truck on a guarded parking ramp, Jen completed the trip to school through an enclosed elevated walkway that deposited her on the second floor, just outside the library. For fresh air, Jen stepped into an outdoor courtyard, enclosed on three sides by the school. When Jen did venture off school property, it was mainly to visit the textbook store across the street. Even at home—Jen and Christine lived in an apartment on the edge of Newark—she erred on the side of caution and rarely ventured into the neighborhood. Jen had no concept of the Newark that existed along Muhammad Ali Avenue and Udele's reticence prevented her from sharing the experience of growing up in the neighborhood where she and Jen now both attended school.

Lest Jen delude herself into believing Newark posed no more a safety threat than Little Egg Harbor, she needed only notice the reminders to the contrary. Like the security cameras strategically placed every few feet on the parking ramp walkway or the two sentry posts mounted, prisonlike, on the school's roof, remnants of the fledgling days of its existence when NJMS was, literally, an institution under siege.

While the campus where Jennifer Hannum pursued her medical degree may not have been as bucolic as, say, 150 acres in Morris County, it was far safer than in the era, shortly after NJMS opened its doors, when one of the school administrators was accosted by a brick-wielding thug. The mugger gave the administrator the option of trading his briefcase and watch in exchange for sparing him a beating with the brick. The administrator handed over the briefcase and watch; the mugger whacked him with the brick anyway, resulting in a visit to the University Hospital emergency room.

Personal safety wasn't all that was vulnerable in the aftermath of NJMS's arrival in the Central Ward. During one memorable period, nearly every television set installed at University Hospital managed to disappear. Police broke up the theft ring—Central Ward residents hired as orderlies had been sneaking the televisions out in laundry carts—and proudly paraded the suspects before the news media. Later, when a group of cops were fingered for stealing computers from the medical school, the incident received minimal publicity.

Eventually, by taking extraordinary safety precautions—slipping past the guards of the medical school without an identification badge became an impossibility—NJMS managed to insulate its students and its property from the world just beyond the school's doors. To offset its image as an academic fortress out of touch with the neighborhood, the school fortified community outreach efforts by increasing the number of free clinics, health fairs and other programs designed to bolster its public image. The gestures, unfortunately, did nothing to appease Central Ward residents with long memories or preconceived notions of NJMS as an island unto itself.

Sitting in the waiting room with Ivan, Jen looked around and began to feel frightened. In the ER, the layer of protection she'd become accustomed to was gone; nowhere were the NJMS security guards who normally stood between Jen and the people of Newark. Next to Jen, a woman sipped from a sixteen-ounce can of malt liquor wrapped in a brown paper bag.

"This ain't mine, it's my husband's," she explained, warily eyeing the ID hanging from Jen's neck. Outside, the temperatures hovered in the teens; obviously, most of the people in the ER were not there to see a doctor.

"Nurse?" the woman asked.

"Medical student," Jen replied.

The woman moved closer and Jen felt her discomfort level rise. This was not what she'd envisioned when she'd set her sights on a career in the ER. As minutes turned into hours, Jen became increasingly agitated over losing an entire night of studying. Finally, at eleven o'clock, she approached the nurse in charge.

"Look, we're medical students here. Give us a pair of crutches and we'll bring them back, he can get an MRI later," she said. The nurse, happy to reduce the throng by two, readily obliged.

By later, Jen meant the next day and not the week that passed before Ivan finally had the knee checked out, a delay brought on by a primary care physician summoned to jury duty. "Welcome to managed care," said Sherry.

Lacking a professional opinion beyond that provided by his lab partners, who correctly diagnosed the injury as a torn medial patellar reitnaculum, Ivan spent the weekend self-medicating with extra-strength nonprescription painkillers, ineffective measures that caused the blood to clot in the knee and his temperature to soar to 102 degrees. When Ivan at last visited the physician, he departed with prescriptions for a stultifying amalgam of painkillers, anticoagulants and antibiotics along with strict orders to stay off the injured leg, a problematic request given his teammates' assumption that he would bear the brunt of the load when they flipped their cadaver.

Not wishing to shirk his responsibility, Ivan put aside his crutches.

Opposite Udele at the shoulder, with Sherry and Jen grabbing hold of the feet, Ivan gamely stepped up to uphold his end of the task, such as it was.

"Clockwise or counterclockwise?" Sherry wondered.

"Counter," said Udele.

"He's been on his back for a long time, I'm afraid to see what it looks like back there," Sherry muttered.

"Formaldehyde bedsores," said Leslie Pooser with repulsively accurate foresight: Up to twenty-four months spent in the supine position had settled the embalming fluid into the rear extremities of the cadavers, causing their backsides to assume the shape of the stainless-steel beds on which they'd lain. Squared off from shoulders to buttocks, the cadavers' backs resembled Naugahyde sofas, circa the 1950s.

"One . . . two . . . three," Sherry counted, and with a grunt from Jen, Sherry and Udele and a grimace from Ivan, the cadaver levitated slightly and, counterclockwise, went facedown, landing back on the table with a slight thump.

"I didn't realize he'd be this heavy," said Jen.

Drawn to a surgical scar midway down the back, Sherry pointed out the tissue had hardened underneath. "He has a Harrington rod," she predicted, using the formal name of the surgical implant used to straighten the spine.

Sherry plucked the purple marker off the table to outline the path of the primary incision on the cadaver's back. Unlike two weeks before, when Jen had been slightly aghast over marking the body in ink, no one said a word.

This time, Udele deferred the first cut to Jen, who warned before taking the scalpel that Ivan wasn't the only one operating in a medicated haze. Taking Sudafed to combat the symptoms of a cold, Jen worried she was too wired to cut, a concern assuaged once she settled into an easy rhythm: After two weeks spent on the body's anterior, the posterior presented a fresh challenge of new and uncharted territory.

Thick and hidebound, the outer layer of skin on the back—the class evolutionists theorized that prehistoric *Homo sapiens* had developed a durable, tough epidermal layer as insulation against the cold—caused the table to go through a week's worth of scalpel blades in barely an afternoon. "If we were real surgeons, we'd be throwing stuff right now, screaming at the nurses to bring us some real equipment," Sherry said.

The impediments presented by the brittle blades notwithstanding, Sherry, Udele, Jen and Ivan approached the upper muscles of the back with a surety that hadn't been evident during the initial dissection of the chest. Resolve may have supplanted hesitation, but the ardor that had been demonstrated from the start never wavered as every revealed muscle, tendon and nerve solicited from Jen, and usually Sherry, an exclamation of astonishment of how closely the structures duplicated the blueprint articulated in *Netter's Atlas.*

Working in tandem on the broad expanse of the back allowed Table 26 to keep pace for the first time with the rest of the lab. No less diligent in reviewing every structure, they somehow maintained a running conversation that darted from the sublime to the trifling: Verbal pinball, a trenchant discussion that one instant revolved around the superior posterior serratus muscle ricocheted into a bitch session about the inequity of the previous Friday's physiology exam. That conversation, in turn, segued into the impact of gross anatomy on shampooing habits, which, somehow, became a discourse on Sherry's study habits.

Sherry's four children had started to refer to her, not necessarily endearingly, as "the mortician," the least of the emotional tariffs that med school in general and GA in particular had exacted on their mother. Two weeks into the semester, the toll of Sherry's time-management predicament could be measured by her impatience with the relative freedom of her much-younger lab partners, frustration that boiled over during an otherwise banal gabfest when Jen, Udele and Ivan reported they'd taken the afternoon off following the physiology exam.

"What a luxury, to go home, blow a day off. I don't have that luxury," Sherry steamed. After the test, Sherry had headed directly to the library to study. At home, Friday night was family night. And despite her best efforts to squeeze in a few hours of studying over weekends, Saturday and Sunday had pretty much been given over to Jerome and the kids, too.

That irony of comparing her life with that of classmates nearly twenty years younger was not lost on Sherry, who, in her first incarnation as a student, had also been chronologically disadvantaged.

On the basis of her performance on a preschool aptitude test, Sherry entered kindergarten as a four-year-old and, thanks to skipping a number of grades en route, began high school at the age of thirteen. Three years later, barely sixteen, she began pursuing an undergraduate degree at Adelphi College on Long Island, a ninety-minute drive from her hometown, Demerest, in north suburban New Jersey.

Unfazed by far older classmates, Sherry viewed Adelphi as a prelude, a stepping-stone to medical school. She had no reason to believe that college would be anything other than an extension of the straight-*A* excursion through high school. At Adelphi, Sherry neither struggled nor excelled her freshman year, compiling a grade point average, 3.0, acceptable to all but someone accustomed to perfection. Ominously, the GPA conspired to alter the course of Sherry's life.

At the conclusion of the spring semester, Sherry met with an academic adviser who recommended she switch her major from premed to nursing. The advice jibed with the oft-preached admonition of Sherry's mother, who felt her daughter lacked the resolve to become a doctor. It was the mid-seventies, and with feminism just taking hold, lessons about unlimited possibility had not yet inculcated most sixteen-year-olds, even those already in college. Buying into the underlying message of medicine being the domain of men and nursing the domain of women, she changed her major.

Three years later Sherry, not yet nineteen, had a college degree and a full-time job in the emergency room at Manhattan's St. Luke's Presbyterian Hospital. A teenager thrust into an adult world, Sherry rejected the frivolity of youth, politely declining invitations to join them at parties or on the flourishing disco circuit, never daring to articulate the lessons of her life experience: "Don't you know people are suffering?"

Young gay men, barely older than Sherry herself, were among those suffering the most in the early days of her fledgling career as a nurse. Flocking to the ER in droves, they arrived at the hospital seeking relief from a disease that didn't yet have a name, a disease that manifested itself with night sweats, pneumonia and Kaposi's sarcoma, a rare epidermal cancer indigenous to men of African descent. Their bodies wasting away, the worst cases required the ER physicians to suction sputum from their lungs. Afterward, Sherry would bathe them, cleansing away the putrid detritus associated with whatever it was that was destroying their bodies.

One night, a man accompanying a patient showed Sherry a picture of a robust, good-looking young man. "A beautiful boy," Sherry remarked, and the companion nodded toward the gurney. Taken aback, Sherry realized the young man in the picture and the patient were one and the same. There was no resemblance whatsoever. Eventually, the National Centers for Disease Control (CDC) came up with a name for the disease that presented itself to Sherry each day. They called it Gay-Related Immune Deficiency (GRID).

Acting on whimsy consciously avoided during the headlong dash through high school and college, in the early 1980s Sherry moved across the country, taking a job at the UCLA Medical Center, where, she discovered, patients exhibiting GRID symptoms were also flooding the emergency room. Not long after she arrived in California, the acronym on the chart changed as a disease with an existence previously familiar only to gay men and health officials began creeping into the collective conscious. Soon, the world would know of AIDS.

Her entrenchment in the front line of an unfolding epidemic notwithstanding, California was a great place to be young and a nurse. By night, Sherry played her part in the never-ending drama scripted by L.A.'s mean streets. Daytime found her on the beach, enjoying the year-round sunshine and balmy weather. Still, even in her version of paradise, Sherry couldn't shed the nagging feeling that she wasn't meant to work rotations on a nursing shift; not a day went by when she didn't think that, perhaps, her time might be better spent in medical school.

The actual road to becoming a doctor began on another whim, an application she dispatched to Columbia University's graduate nursing program. Columbia was tough, among the top nursing schools in the nation, and Sherry thought she didn't stand a chance. To her surprise, the letter Columbia sent told her she'd been accepted.

Following two years of grad school, Sherry planned to spend a year working to save money while simultaneously applying to medical schools. She got the degree, with a specialty in anesthesiology to boot, but a funny thing happened to interfere with part two of her plan: She fell in love.

On a blind date she met Jerome Ikalowych, the small-town, no-nonsense cop with intense brown eyes that connected immediately with Sherry Zilbert. From the outset, the outgoing nurse let it be known that, henceforth, no obstacle was about to get between her and her lifelong ambition to become a physician.

Two years after Sherry married in 1983, an obstacle—albeit a happy one—did, in fact, materialize to impede her path to med school. His name was Jason. Five years later, when Jason reached school age, the door again inched open only to slam shut when Sherry learned another child was on the way. With Sarah's birth, Sherry assumed she was done making babies and began preparations for medical school by enrolling in the prerequisite and refresher courses that could sway an admissions committee to accept a nontraditional student.

Unexpectedly pregnant in 1993, with twins no less, Sherry refused to allow circumstances to further derail intent. On May 4, 1993, twelve hours after she aced a biochemistry final exam, Sherry went into labor. The twins were born via C-section the next day.

And that, Sherry explained to Jen and Udele as they began dissecting the intermediate muscles of the back, is how she came to become a first-year medical student at the age of thirty-nine. "If I were a man," she said, recalling the Adelphi academic counselor who sidetracked her twenty years earlier, "they would have pulled me aside and said, 'Pull your grades up.' But no one said it to me. Girls didn't go to medical school. To me, they said, 'You can't go to medical school, how will you have a family?'"

Udele and Jen nodded their heads in assent, retaining what Sherry was saying intellectually in much the same manner they comprehended the disco era, American involvement in Southeast Asia or a time when women were expected to sacrifice personal ambition at the altar of motherhood. The short chronicle of Sherry's sojourn to med school smacked of not-quite-ancient history to Jen and Udele, two of seventy-one women in the Class of 2002, each culturally inculcated from the time they were little girls with the belief that there was nothing in the world they couldn't do.

Empowerment at Table 26 translated into the women assuming

that, crutches and all, Ivan would continue to perform all functions that didn't involve a scalpel and forceps, including the "honor" of using a stainless-steel hammer and chisel to execute the laminectomy, medical jargon for severing the spine. As the dissection proceeded through the deep layer of muscles of the back, it became clear that Sherry, till then batting a perfect 1.000 in the prediction department, had been dead wrong about the presence of a Harrington rod.

The misdiagnosis suited everyone just fine, especially Ivan, who ascertained that the laminectomy, even without a surgically generated anomaly, would pose enough problems. Taking seriously the faculty warning that embalming fluid, spinal fluid and other unexpunged body fluids pooled near the base of the spine held the possibility of turning the laminectomy into a rather messy affair, Ivan accepted Sherry's offer of a surgical mask and goggles before striking the first blow.

The number of students at the sink rinsing bone chips and liquids of indecipherable origin from their faces made Ivan grateful for his lab partner's largesse. Advising that it would be a "tedious but rewarding process," the syllabus's assessment of the laminectomy turned out to be most prophetic.

Some found the rhythmic cadence of stainless-steel hammers simultaneously striking stainless-steel chisels oddly comforting, likening it to a sound they associated with Santa's workshop. Others, notably Udele, flinched and turned away, visibly unsettled by the force necessary to cut through the vertebral canal and spinal cord.

While the sinewy resilience of the spinal column indeed proved formidable in its resistance to the hammer and chisel, that alone was not the only source of Ivan's vigorous application. That morning the scores of the latest physiology exam had been posted, with a mean score of seventy-five. "And I was way below that," Ivan reported to Jen after they'd arrived in lab. He shook his head sadly. "You don't even want to know how far below."

Jen said nothing, joined in her quietude by Udele and Sherry. Discussing one's grade in med school is a little like talking about one's salary in the workplace: It's generally considered impolite to say and impolitic to ask. Sensing that among his lab partners he alone had the need to grouse about a grade, Ivan dropped the subject.

But clearly, the test score continued to rankle. And the more Ivan hammered, the more his agitation demonstrated itself. Fortunately, the laminectomy provided ample opportunity to expend frustration. While his lab partners complained about NJMS's decision to schedule physiology and gross anatomy in the same semester ("When I told a friend of mine at another school, he couldn't believe they're doing this to us," said Sherry) and their auspicious entry into medicine during the advent of managed care ("It can't be about money anymore, it has to be about medicine," Jen, ever the purist, declared), Ivan kept chipping away at the spine, stopping every so often to wipe his brow.

Finally, a full hour into the laminectomy, Ivan victoriously thrust hammer and chisel skyward, signaling the job was completed. Coming as it did five minutes before the weekend break, his timing could not have been better. Departing from the pattern of painstaking review, the table quickly went over the pertinent structures, lingering only to consider what occurs when, as the syllabus put it, "the prolapse of the nucleus pulposus through the outer annulus fibrosus, [causes] impingement on nerve roots exiting the spinal canal." Satisfied that she understood what occurs when a disc slips, Jen excused herself and headed for a basketball game, ATP/2002's first without Ivan.

Sherry was bound for a sporting event as well, but before leaving for Jason's wrestling match, she lingered after the others had departed to attend to some motherly nurturing. From her backpack, Sherry produced two pairs of worn tube socks. One pair she carefully slipped over the cadaver's feet, the other she placed around his hands. Next, she wrapped the four socks in sandwich bags secured by rubber bands. Satisfied with her handiwork—"Necrosis, cell death. Causes the skin to wither, atrophy, blacken," she said quietly, before slinging the backpack over her shoulder and making her way to the door.

Arriving just as Jason's junior varsity wrestling team took the floor, Sherry caught her son's eye before taking a seat halfway up the bleachers, where she discreetly removed her notes and *Snell's Clinical Anatomy* from the backpack. Out of necessity, she'd become expert at multitasking, taking particular pride in her ability to simultaneously watch a wrestling match while absorbing the minutiae of physiology and gross anatomy.

The trick was to segregate herself from the rest of the crowd, a tactic that gave her a measure of privacy while sparing others in attendance exposure to the first-year smell. Fans who couldn't take the hint often paid the price in more ways than an assault on the olfactory nerve. The week before, a gaggle of teenage girls grabbed the seats behind Sherry and were giggling their way through the meet when one of them stole a look at the oversized textbook in Sherry's hands. Sherry heard a gasp, more giggles and then the sound of feet rushing to put distance between themselves and the woman lost in an anatomy atlas illustrated with photographs of cadavers.

Jason sometimes chastised Sherry for keeping her head buried in books while his teammates competed. "I watch you, do I have to watch the other kids, too?" Sherry asked, knowing full well the answer.

Had the second semester not instantly become a cliché—for Sherry, there literally were not enough hours in the day—she could have dwelled more not only on Jason, but Sarah, Paul, Stephen and Jerome as well. "At the very least, [school] is torturing my husband and inconveniencing my family," she said, understanding that, practically speaking, once the process had been set in motion she could no more stop it than tack extra hours onto the clock.

Instead of losing the battle to stave off sleep every night, she wished she still possessed the stamina of Udele, Jen and Ivan, able to bounce back bright-eyed after studying until two or three o'clock in the morning.

Still, a day didn't pass that Sherry wasn't grateful she was no longer twenty-two years old. Noticing that some of her classmates had started to pair off caused Sherry to recall the tenuous nature of new romantic entanglements, and made her all the more grateful for the stability that came from having a home, a husband, a family.

Sherry saw that she brought to med school a sense of self that couldn't reasonably be expected from kids fresh out of college. She'd seen the pain of others and she'd suffered pain of her own, notably two miscarriages that she and Jerome had endured. Sherry wasn't so old she didn't recall that, in addition to a relatively carefree existence, twenty-something meant dealing with a lot of extraneous distractions. And though it might have taken a thirty-nine-year-old mind a bit longer to grasp what was readily apparent to most of her class-mates, Sherry was satisfied knowing the preoccupations of young adulthood were behind her.

Except in the massive amount of free time available to them, rarely did Sherry view her lab partners with envy. She was nearly forty and they were not. That's the way it was. Embracing the age discrepancy served her well in an atmosphere that offered not a moment for intro-spection or regret. The pressures of the second semester were such that dire consequences awaited those who didn't keep the focus on what lay immediately ahead. Which, in Sherry's case, meant not the outcome of a high school wrestling meet but an onslaught of exams just over the horizon.

For the sake of appearance, Sherry looked up occasionally from her textbooks to see what was transpiring on the wrestling mat. Still, only when Jason wrestled in the middleweight division did she give the competition her full attention. After he won, Sherry gave Jason a smile of proud approval, holding her son's attention until his team-mates rallied around him in congratulations. Her maternal obligation fulfilled, Sherry Ikalowych returned to the gross anatomy notes sprawled on the seat before her.

CHAPTER 9

———————•———————

A s one who had spent an above-average amount of time listening to the roar of crowds, Jen couldn't imagine studying amid the din of rabid wrestling fans. When Jen studied, she liked it quiet and, paradoxically for someone who so enthusiastically embraced the concept of teamwork, she liked going it alone.

During the first semester she'd made a stab at being sociable, accepting invitations to be part of the study groups, often joining her classmates afterward for beers at local pubs, all the time never feeling as though she were part of the crowd. Class distinction played a role: Although Jen didn't live frugally, she was definitely restricted to a budget dictated by the amount of her student loan. Compared with that of George Washington University and the Ivy League schools, the in-state tuition at the New Jersey Medical School, $16,500, was a relative bargain. Still, when books, rent, food and transportation were added to the flat cost of her education, by the start of the second semester Jen was already $30,000 in debt.

From the hand-me-down luxury cars given them by their parents—in contrast to Jen's inherited vehicle, a dilapidated Ford Ranger pickup that once belonged to her father—to the way they blithely dropped twenty-five bucks on meals, it didn't take long for Jen to conclude that many of her classmates had emerged from a financial stratosphere far beyond that provided by a fireman and his wife. At NJMS, the lack of dormitories or a dominant student ghetto made dropping from the social circle relatively easy. Scattered throughout Essex, Union and Hudson counties, the Class of 2002 lived no place in particular. In every sense a commuter school, at the conclusion of the academic day the students hopped into their cars and went their own way.

Jen went farther than most, preferring to study at a Barnes &

Noble bookstore in the nearby suburbs than the apartment she shared with Christine Ortiz. Part library, part hangout, the Barnes & Noble appealed to Jen primarily as a setting devoid of individuals obsessed with the nuances of medical school.

Conversely, the environment Jen so mightily distanced herself from every night had, in effect, become a self-contained universe for one of her lab partners. Udele practically lived at school, arriving around nine in the morning for her first class and often not departing until well past midnight, sealed off nearly fifteen hours every day from outside air and light.

Fulfilling her father's adage—"Learning is depressing, lonely, exhausting"—Udele usually studied alone. In the first term she had studied in tandem with Joyce Prophete, one of five women of African or Caribbean descent in the class. Joyce shared with Udele a similar temperament: Both quiet, intense, focused and difficult to approach, they allowed few into their closed circle.

Udele protected her privacy to the extent that, in an atmosphere overflowing with verbal and sartorial references to undergraduate affiliations—sweatshirts trumpeting the names of alma maters became practically de rigueur once the weather turned cold—it wasn't until the second semester that she learned another member of the Duke class of 1998 had joined her at NJMS.

Eventually, even Joyce's quiet company became too distracting, and Udele began studying on her own. Never a social butterfly, Udele withdrew further in med school, closing out everyone, even her own family. "I miss you, I feel like I'm losing you," Chris Tagoe told her daughter one morning as Udele scampered out the door. In a way, she already had: Udele's life revolved so completely around anatomy and physiology that lectures and lab became, in effect, a diversion from studying. Even in lab she often appeared preoccupied and distracted, as if consumed by guilt that a few hours were allowed to pass without submerging herself in a textbook.

Udele studied wherever solitude happened to draw her: sometimes one of the study rooms sprinkled throughout the building's first floor, more often in library cubicles where the table graffiti reflected how medical schools differ from other institutions of higher learning. In the George F. Smith Library of the Health Sciences, pronouncements detailing the technical expertise of individuals willing to share a "good time" were far outnumbered by admonitions—"Sleep Well, Study Hard, Play Hard and You'll Make It"—and laments—"No matter how hard I study, I can't do well in my courses." The latter snippet prompted responses ranging from "Me, too!" to "So you are dum *and* stupid" to "Be Positive!" Dominating all other musings etched into various crannies throughout the library were formulas decipherable not even to the intended audience, case in point being the succinct response scribbled under a notation reading "=1/2 = .0693/KE; CO=SV (over) HR and ex=clear + plasma cent": "Huh?"

Aware of how she'd allowed school to dominate her, Udele did not know how to control it. "Sometimes I wonder why this is the one thing that is constant in my life," she said one day before lab. "I just know I have to do this. Some people have other things in their lives, they go out after school with their friends and they go to parties. I like to go out, too. But I don't, not usually. This always has to come first."

The dynamic of the gross anatomy lab, a situation where shared responsibility might have alleviated some of the pressure, only exacerbated Udele's self-imposed regimen. Superficially, the reason she'd assumed responsibility for shepherding Table 26 through the dissection had to do with her experience in the summer program. Circumspectly, Udele understood her need to control and monitor every aspect of the table's lab work stemmed from the same compulsion—perfectionism—that kept her at school all day and all night. As is the case with most perfectionists, the locus was rooted not in a sense of superiority but, rather, quite the opposite. Rare was the day that Udele didn't ask herself, "Am I smart enough to do this?"

Unlike Jen, Ivan and Sherry, who filled the time when they weren't cutting with idle conversation, Udele remained riveted to the task even after relinquishing the scalpel and forceps. Sometimes, like the afternoon when Udele attempted to prevent a dissecting error by compulsively thrusting a hand between the scalpel and a back muscle Jen was about to obliterate, vigilance nearly resulted in disaster.

Concerned that her enthusiasm for dissecting might be diminished the second time around, Udele's zeal was instead enhanced by familiarity with the anatomical terrain. So much so, in fact, that she feared Jen, Ivan and Sherry were becoming resentful of the way she dominated the process.

Although on a personal level Udele may not have felt particularly close to her lab partners, the last thing she wanted was to upset the dynamic of a table she thought worked well together. Udele, the perfectionist, recognized she couldn't have been assigned a better table than the one that included the fastidious Sherry and Jen, unrelenting in their insistence that every structure be reviewed until the function was understood by all.

Beyond an appreciation for the different talents her lab partners brought to Table 26, Udele was also becoming rather fond of the man on the table. Sure, the cadaver had some girth on him, but to Udele's way of thinking, that gave him a measure of character absent in the scrawnier cadavers: "All the others look the same, they're all so small. Our cadaver is big, a big man. He has muscles. The others have muscles that are one color, a monotone. Maybe he played sports when he was younger and then, in older age, became less active. Whoever he was, it's like Sherry says, he sure seemed to enjoy himself a lot."

Understanding that the safest emotional route was to regard the cadaver only in terms of his anatomy, Udele struggled not to consider the manner or means by which the man found enjoyment during his life. Still, having well absorbed the overarching lesson of problem-based learning—"Unconditional Positive Regard"—Udele was powerless in resisting the tug of empathy as the cadaver's history of respiratory and cardiac problems came to the fore. Often, she'd find herself wondering not so much about who he was but, strangely, what his family thought of him being there, with them, doing what they were doing to his body. Always, introspection was fleeting; the work that needed to be done left no choice but to again establish the necessary distance between herself and the cadaver.

Throughout lab, the first signs of emotional withdrawal occurred subconsciously the afternoon the lungs were extracted and rinsed in an ordinary sink. With the attachment of the air hose to the organs, causing them to expand and contract like footballs, the capacity for shock began to diminish. From that point forward it became foreordained that acts family and friends might find horrific would become commonplace. With that inevitability, the people lying atop stainless-steel tables began to gradually pass from the realm of human being to that of case study.

In the weeks that followed, self-preservation created an emotional shield as the students necessarily relegated the donors to a province other than humanity, a place where—for fourteen weeks, anyway—they became unique only in relation to the integrated system that had once been the source of their existence.

And yet the students could never manage to completely tuck the spirit of the cadavers into that corner of the mind that made it possible to forget that somewhere these men and women were cherished for what they'd been in life, not the lessons they were imparting in death. Introducing themselves into the lives of their dissectors by smell and touch, the cadavers became part of the students' existence, insinuating themselves into ordinary conversation and, despite Anthony Boccabella's findings to the contrary, extraordinary dreams that lasted the entire semester. Among the students, the cadavers became fodder for clinical observations at lunch and parties; for the students' friends and families the bodies were objects of morbid fascination:

"This is Jen," a friend of her brother's said one night as he introduced her to a roomful of strangers. "She cuts up human bodies."

Equally tantalized by what they were doing in lab, the fascination of the students assumed a different construction, one far more pertinent to the experience. Prior to lab one afternoon, upward of fifteen male students were crammed into a rest room changing from street clothes into scrubs (of the thousands of doctors who have matriculated through the New Jersey Medical School, not one had bothered to endow his or her alma mater with locker room facilities for first-year students) when the banter turned to whether the anatomy department imposed any standards before accepting the bodies of donors.

"Don't think so, but they ought to," someone shouted from behind the locked door of a toilet stall.

"Yeah, what they should do is make them fast for the last ten days of their lives so we're not steeped in shit when we get to the abdomen," agreed a second anonymous voice.

"Either that, or force them to have a voluntary enema forty-eight hours before they die," said another.

The students had no way of knowing the donors themselves often raise the same issue (although, to date, none have gone so far as to volunteer for a deathbed enema). One such gentleman once turned up in the anatomy department office to announce that after months of resistance, his skeptical son had finally been persuaded to sign on to his father's desire to bequeath his body to science. "I'm here to interview to become a cadaver. . . . I have had only one operation, my appendix is missing, and other than that I'm in great shape: I eat right, I don't smoke and I exercise every day," the old man proclaimed, receiving the necessary papers from Essie Feldman.

Had a donor made an appearance before the students before they began dissecting, it might have put to rest the overriding question about why individuals donate their bodies. Furthermore, seeing a donor in person would have cleared up the misconception, advanced by those who weren't paying attention during the portion of the introductory lecture when the procedures for procuring bodies were explained, that the men and women on the tables were not indigents making a stop en route to a potter's field.

Whereas it was once common to appropriate the bodies of the destitute and downtrodden, today's donor programs, such as the one in place at NJMS, have resulted in a surplus of cadavers, available to medical schools across the country. A notable exception to using the bodies of donors to further medical education is the Albert Einstein College of Medicine in the Bronx.

Operating under the auspices of Yeshiva University, which holds to strict tenets of Jewish Orthodoxy, Albert Einstein is restricted by religious law from actively soliciting body donations. Should an individual wish to donate his or her body, Einstein will eagerly accept it, but because the school hasn't a procurement program similar to those at secular universities, nearly two-thirds of its cadavers are secured from the office of the medical examiner.

Knowing an individual did not volunteer his or her body for the purposes of dissection sometimes weighs heavily on first-year students enrolled at the Bronx campus. Few experience the guilt of David Palencia, a former marine who, after abandoning a career as a graphic designer to enter medical school at the age of thirty-four, seriously contemplated walking away rather than proceed with an act he came to view as "sacrilegious."

David's predicament arose from a dissection that betrayed more about the cadaver's station in life than it did about his anatomy. The extent of pathological evidence—poor diet, inadequate dental care, hygienic lapses, nicotine and possibly alcohol addiction—unsettled David to such distraction that "I could not find the spiritual or intellectual relevance to dismember a body. Especially a body that I believe to be unclaimed. I don't believe this body was donated by this gentleman before he died or by his relatives after his death. As a member of an ethnic minority [Hispanic], and a distinct minority at this school, I didn't feel it was right to dissect the body of another ethnic minority [African-American]. If this had been someone who had given his body freely to help us learn medicine it would have been one thing, but being that this is an unclaimed body, it didn't seem right. It seemed like we were defaming the dead, dismembering someone who loved, who was loved, who breathed and wanted things. I'm not sure this was what he wanted."

On the cusp of giving up his quest to become a physician, David Palencia confided his anguish to a friend, a painter and, "I know it sounds like a cliché, but a venerable old Chinese guy." The friend went with David one night to the lab, where together they studied the product of the dissection. David showed him the muscles and the bones, the arteries and the nerves, the organs and tissues. The student told the old man that no matter how much his dissection team labored "not to destroy structures, but to see structures," he could not shake the notion that they were violating the body.

The old man took it in and, in a measured voice, imparted to David a smidgen of wisdom: "As a painter, I cannot learn to draw just by looking at human models, I have to look beneath the skin. And you have to look at him beneath the skin, too, so that you can understand."

The advice had the desired effect: "After that, I stopped looking at the cadaver as a victim and began looking at him as my teacher," David recalled.

Given the medical establishment's checkered history in securing bodies in the name of anatomical research and education, David Palencia was fortunate in that he came to medicine during an age of comparative enlightenment.

Dissecting bodies as a means of comprehension is a relatively recent phenomenon dating to Andreas Vesalius (1514–64), the father of modern anatomical studies.

The son of a prominent Belgian pharmacist, as a child Vesalius dissected animals to compensate for the lack of adequate books detailing the anatomy. Educated at both the University of Leuven and the University of Paris, in 1537, Vesalius wound up at Padua (Italy), the eminent medical school of the time. A day after graduation, at age twenty-four, he was appointed a professor. By then, Vesalius had already written a paper rejecting anatomical guidelines that had been in place since Galen's publication, fifteen hundred years before, of *On the Uses of the Parts of the Body of Man.*

To Vesalius, the obvious flaw with *Body of Man* was disingenuity: Galen's signature template was predicated not on man but animals. In his own research, Vesalius did not make the same mistake, basing his every pronouncement on personal observation, not theory or theology. The landmark lectures Vesalius delivered at Padua included the use of the outer anatomy of live humans for purposes of demonstration. By 1537, the models brought by Vesalius to the amphitheater were no longer alive.

Ultimately, the deceased criminals and various other unfortunates extricated from the morgue before they could be carted off to a pauper's grave provided the groundwork for his seven-volume epic, *De Humani Corporis Fabrica* (On the Workings of the Human Body), still considered the definitive anatomical blueprint.

Supplemented by the artistic brilliance of portraitists Titian (the Venetian, Tiziano Vecellio) and Jan Van Calcar, whose illustrations instigated a lineage of templates that eventually extended to *Netter's Atlas* as well as *Grant's Dissector,* Vesalius opened a door that had been closed for time immemorial. Through it poured hundreds, and eventually thousands, of anatomists and physicians.

Incorporating physiology into the anatomical form, Englishman William Harvey (1578–1657) became the next key figure to add to the body of work. The application of physiology—"each organ has a dis-coverable function and is related in its mode of working to all other organs and the body as a whole"—empowered Harvey to reject the medical establishment's Galenic model of the flat-earth theory. Blood, declared Harvey, circulates in one direction.

As anatomy became synonymous with medical science, the demand for cadavers began to exceed supply, especially in the academ-ically charged atmosphere of late-nineteenth-century England. In a country where the leading teacher/surgeons, many self-appointed deans of their own private academies, competed ferociously for top-drawer students, the combination of research and education proved too much for the established system for securing bodies. In other words, there weren't enough dead criminals to go around.

Into the breach on the behalf of one Dr. Robert Knox of the Royal Academy of Surgeons stepped William Hare and William Burke. Hare and Burke's contribution to medical science, such as it was, is the stuff of legend. Depending on who is telling the story, Hare and Burke started their careers as body-procurers in the fairly standard manner of the day: They robbed from graves. When this mode proved insuffi-cient to meet the needs of their customers, the pair progressed to murder.

The pair's modus operandi involved enticing a target, usually a lady of the night, to a rooming house where Burke would pinch the victim's nose until she lost consciousness. At that juncture, Hare took over, sitting upon the chest until the life ebbed out of her. Hare and Burke may have been without remorse, but they weren't stupid: Min-imizing the physical damage made the cadavers more attractive to a clientele that preferred its anatomical specimens fresh and relatively unmarred.

Hare and Burke's downfall began when Knox's students, patrons of the Edinburgh night trade, recognized a cadaver. Knox, who professed to know nothing of the manner by which the bodies were acquired, was thrust into a lifetime of disgrace after the Edinburgh anatomy murders were solved by Scotland Yard. He fared better than William Burke. Betrayed by a partner in crime who copped a plea, Burke was subjected to a public hanging, after which his body was publicly dissected by one of Knox's rivals. Thousands attended both the hanging and the dissection. To this day, Burke's skin covers a score of the books housed in the library serving the Royal Academy of Surgeons.

In the annals of nineteenth-century American medical schools, the attainment of bodies never sank to the level of infamy recorded in Edinburgh. Not that the early practice of dissection in this country hadn't its own dark side. As in Europe, the bulk of the bodies delivered to U.S. anatomy labs during the 1800s were those of criminals and indigents; in the aftermath of heinous crimes, the judicial system considered postexecution dissection the "ultimate punishment."

Still, because the volume of dead hooligans never matched the number of viable bodies needed by the burgeoning field of medical education, many cadaver-suppliers resorted to working by night, surreptitiously opening fresh graves. Medical school administrators, failing to heed the lesson of Dr. Knox, were careful not to ask too many questions.

Only due to a macabre coincidence did the practice of grave robbing come to a halt.

Shortly after the 1878 death of U.S. senator John Scott Harrison of Ohio (the son of former president William Henry Harrison and the father of another, Benjamin Harrison), another of John Scott Harrison's sons paid a visit to the Medical College of Ohio. During the informal tour, the younger Harrison was escorted to an anatomy lab, where, to his great distress, he happened upon the partially dissected body of the father he had helped bury just weeks before. The unwanted fate of John Scott Harrison set in motion the wheels of government and, not long afterward, federal and state statutes were enacted to permit the willful donation of bodies to science.

Five score after the laws inspired by the unauthorized procurement of John Scott Harrison's body, the four medical students stationed at Table 26 in the gross anatomy laboratory operated by the University of Medicine and Dentistry New Jersey–New Jersey Medical School spent not an inordinate amount of time considering the life of the man who, because of those long-ago reforms, had voluntarily availed himself of them. Udele, Sherry, Ivan and Jen never gave their cadaver a name and, in fact, never even considered it. Ivan and Jen came closer than the others, occasionally referring to Number 3426 in the third person as "our guy."

At the adjoining table, number 27, Priya Singh, citing what she believed to be an uncanny resemblance to the queen of England, began calling her cadaver Elizabeth. None of Priya's lab partners followed suit (although one, in respect to the cadaver's diminutive stature, suggested they call her Slim). Other than Priya, barely a handful of tables bothered to christen their cadavers, a practice some viewed as a sign of disrespect: "She already had a name, it doesn't seem right for us to give her another," one student explained.

To some, bestowing names seemed a dubious gambit in an environment where each day the struggle began anew to keep the focus on anatomy and not humanity. In that theater of thought, naming a cadaver just made it that much more difficult to pull back.

For her part, Jen never did bother to detach, choosing, instead, to build on the connection that evidenced itself the first day when she defended the cadaver from her classmate's barbed comment about the donor's weight. For reasons not entirely clear—the man, after all, was dead—Jen had a real affinity for him, a regard that grew exponentially as the class progressed.

Broad, with well-articulated musculature, Number 3426 had great shoulders. The best shoulders in Lab C, Udele thought.

"It looks like he was pretty active before his health problems," said Sherry.

"We don't know what he did. Maybe he worked outside, had an active job, an outdoor job. Maybe he was a landscaper or something," said Jen, watching Ivan slowly separate the overlaying muscles wrapped around the shoulder.

Still unsure of himself—the knee injury and a general reluctance to cut had made him less sophisticated than the rest of the table when it came to dissecting—Ivan worked slowly, methodically lifting and scraping extraneous tissue from the muscles.

Only two months before, his exposure to human tissue had been limited to the examination of cells magnified one thousand times on the Olympus microscopes in the cell and tissue biology lab. That had been a common experience, the entire class basically looking at the same sample smeared pristinely across a slide, everyone comparing the same notes, all of it quite mundane.

From the first-semester perspective of CTB, gross anatomy had seemed exotic, fascinating, edgy. And while the first two weeks had been everything Ivan had expected, something changed during the dissection of the back. Dissecting a human body, he began to realize, could often be as tedious as examining slides under a microscope.

"Eighty percent cleaning up, twenty percent glory," he told Jen, putting aside the scalpel.

"We're supposed to cut through the whole thing," she reminded him, pointing to the muscle he'd left half-severed.

Ivan smiled. "I know, but let's leave some of that muscle in there. I like that muscle."

"You're so cute," Jen laughed. Picking up the scalpel, she completed the job Ivan had started after reciting the muscle's formal name twice, imprinting it in her memory. To Jen, the hours spent every night in the company of *Snell's Clinical Anatomy* made little sense until she saw the form and function inside the cadaver for herself, although, sometimes, even viewing a configuration quite literally in the flesh failed to translate into comprehension.

"And what is this?" In the teeth of his forceps, Zolton Spolarics squeezed a mystery tissue jutting prominently from underneath a tendon in the arm. Jen looked at Ivan, Ivan looked at Udele, Udele looked at Sherry. Sherry looked blankly at the instructor.

"Don't know?" Spolarics continued. "Well, let me tell you, this is the most beautiful one I've seen here so far."

Jen caught Ivan's eye and mouthed, "What nerve is it?" Ivan shrugged. Spolarics mumbled an answer, "the left anterior interosseous nerve." Sherry appeared to be the only one to hear him; puzzlement remained on the faces of Ivan and Jen.

"It doesn't look like that in the book; in the book it looks farther away," Sherry noted, with an air of authority.

"Look at the other cadavers, look at"—he pointed to tables 25 and 27—"you won't see it. It won't look like this," said Spolarics.

"Does its prominence have to do with handedness?" Sherry asked.

"Handedness?"

"Yeah, you know, whether he was right-handed or left-handed."

Spolarics considered the question. "It's just the development," he answered absently.

Sherry tugged at the nerve. "Oh, God, this is just amazing," she said as Spolarics, without further comment, walked in the direction of the hand raised at Table 30.

Alarmed at the importance Spolarics had attached to the structure, Jen took the instructor's place next to her lab partner. "Which nerve?" she asked anxiously. "What's the name?"

Sherry exaggerated a shrug. "Damned if I know," she responded, revealing the entire performance had been an act on the behalf of Spolarics. "I have no idea what he was talking about. The only one I can identify now is the musculocutaneous."

Sherry's operating room experience, which typically put her one step ahead of the rest of the table when it came to identifying pertinent tissues, normally served her better when it came to keeping abreast of the department's emphasis on the liaison of anatomy and function. Sometimes, though, the anatomy befuddled even the most seasoned student in lab. As happened again a short time later when Sherry encountered the palmaris longus muscle.

Located in the forearm, the muscle is absent in 14 percent of the population. "But here in this room we have only three who don't have it. Strange," exclaimed Dr. Ajit S. Dhawan, an instructor rotated into the lab to assist Spolarics.

"If they don't have it, then where did it go? Evolution?" asked Jen, turning evolution into a destination.

"Evolution, right. We don't need it anymore," said Dhawan, hurrying away before being confronted with questions extending beyond the realm of an anatomy instructor. In Dhawan's absence, Jen continued to press the issue: "I don't understand, where did it go? If it's not here, then where did it go?"

"I guess we just don't need it anymore," said Sherry.

"Maybe it was used to grab on to trees, to swing like Tarzan. Maybe Tarzan used it to pick up chicks," Ivan volunteered.

"Does that work?" Sherry inquired.

"Swinging from trees to pick up chicks? I wouldn't know. I have a girlfriend," said Ivan.

"Girlfriend?" Jen asked. The existence of a girlfriend constituted news; Ivan shared so few details of his personal life that, after a month together, his three lab partners, Jen included, really knew nothing about who he was or what he'd come from.

Jen tried to turn the conversation back to the business at hand— "Well, if it's something we don't use anymore, maybe there's something else we use that has grown in its place"—but got nowhere. The hour was late, everyone was exhausted and there was already enough to think about without lumping Darwinism into the equation.

Consideration of the somewhat-extinct palmaris longus notwithstanding, the arm ushered in a correlative phase of the dissection missing from the previous procedures. While intellectually they understood that their own thoracic cavities and spines replicated those inside the cadaver, the structures contained a paucity of external reference points. The beat of their hearts, the rush of oxygen filling their lungs and even strain in the upper back—the levator costae muscle—could be considered a product of function rather than form. Consequently, what they encountered as they foraged through the thoracic cavity and along the spine more often resembled the sensation of beholding the components under the hood of a high-performance automobile than what lie under the skin of a human being.

The arms were different, appendages in which they could see the bulge of muscle, the ripple of tendons, the knot at the base of the wrist known as the styloid process of ulna. For the first time, they were able to see themselves in the anatomy. When flexed, the cadaver's exposed muscles expanded exactly as theirs did; the action of the cadaver's elbows and wrists mimicked the function of their joints.

Familiarity emboldened them: No longer was the scalpel the enemy; the act of dissection ceased being a deed that summoned fear. It's the juncture, said Vasan, when the "students slowly understand themselves. They see how [the anatomy] looks and they start to realize that that's what they look like inside. They put themselves in there. It's as though their own bodies were lying before them and that's how they relate the cadaver to real life."

"Watch this," said Sherry. Using the forceps, she tugged at the flexor digitorum, a tendon in the forearm, and the little finger began to wiggle. One by one, she did the same with tendons attached to the other four fingers and the thumb.

"That is cool," said the usually unflappable Udele.

"Yeah, I know. And they're in order, too," Sherry gushed.

"Which one do we use on the [New Jersey] Turnpike?" Jen asked. Sherry went back to the extensor digitorum. "This one," said Sherry, pulling on the tendon, causing the middle finger to jut upward.

"Ah, the New Jersey salute," said Jen to the giggling Udele.

Sherry deflected the table's attention to the extensor policis in the web of the hand between the thumb and forefinger. "When I worked in the ER on Sunday mornings people used to come in all the time with this sliced up. We used to call it the bagel injury. Every Sunday morning someone would come in after slicing through a bagel without getting their other hand out of the way, cutting the ulnar nerve," she explained.

Jen hesitated before correcting her. "Recurrent branch of the recurrent meningeal nerve," she said quietly.

"Recurrent? You sure?"

"Recurrent," said Jen firmly.

"Right, right. Median recurrent. They'd slice through the bagel, through the muscle, through the recurrent nerve, through the tendon and that was usually it for them as far as ever moving the thumb again."

As comfortable with one another as they were with the routine, Table 26 sashayed through the dissection of the hand, blissfully unaware that in other quarters the procedure was one that again capsized the emotional equilibrium. The hand, in fact, made some students so uncomfortable that they deferred the dissection to their lab partners.

"The first contact you have with another person is with his or her hand. It's a sign of humanity," explained a student who won a bout with trepidation before she was able to assist her team in the dissection of the hand. Meanwhile, a student at another table not only refrained from dissecting the hand, she couldn't even watch as her lab partners moved through the procedure.

Barely a year before, the student had sat, day after day, at her grandfather's bedside, holding his hand, offering comfort as, slowly, his life slipped away.

"He was completely like this . . . exactly like this," the student said, gesturing toward but careful not to touch the outstretched palms of the prostrate cadaver. "Exactly."

Opposite those distressed by the hand and its dissection were students like Jen. Jen was a hand-holder; leaning over to inspect a structure, she'd subliminally slip a gloved hand into Number 3426's palm, keeping it there interminably if she happened to be engrossed by whatever was transpiring at that point of the dissection.

When once someone pointed out to her that she was holding hands with a cadaver, Jen serenely proclaimed herself not in the least embarrassed. Unaffected, she kept her hand right where it was, as though it were the most natural thing in the world.

Before they were destroyed by forceps and scalpel, Jen got to know the cadaver's hands very well. So well, that from their texture she amended her earlier hypothesis, based on the musculature of his shoulders, on how the donor might have earned a living.

"He's a carpenter," Jen declared.

"He changes professions every day," Ivan protested, reminding Jen that not twenty-four hours before she'd concluded he'd earned a living as a landscaper.

"I know, but look at these hands," Jen said. "These are the hands of a carpenter. I'm sticking with that."

With Jen's latest proclamation, the table tucked the cadaver's humanity back into the safe place. Without a doubt, it would emerge again the next week with the advent of the most electrifying dissection of all, the face and head. The Class of 2002 knew the time to obsess over what they were about to do to the essence of the cadavers' identity would come soon enough. But first they had a rendezvous with the seventh circle of medical school hell.

CHAPTER 10

———————•———————

Had Nagaswami Vasan his druthers, the New Jersey Medical School would not have yoked physiology and gross anatomy in the same semester. Nor would he have put the class through two major exams in three days, a cruel subjection the course coordinator blamed on a physiology department "jealous of anatomy because anatomy is more exciting. If they think anatomy gets too much attention, well, that's not my fault."

Ever the student's professor, Vasan was wholly sympathetic to the medical school equivalent of piling on, a physiology test on Friday followed by the bifurcated written and practical gross anatomy exam on Monday. The solution to stacking one exam atop the other, he knew, was to abandon the notion that gross anatomy was a course best offered after the students had the opportunity to acclimate to the medical school environment, a policy that each year crammed the two toughest first-year courses—GA and physiology—into the winter term. (Indeed, after a decade of offering gross anatomy during the second semester, during the 1999–2000 academic year NJMS joined nearly every other medical school by switching the course to the fall term so it no longer conflicted with physiology.)

Still, Vasan's empathy for the students only went so far. After all, his own preparation began so long before the exam that, by the time the first incision had been struck in lab, the textbooks and anatomical tracts accumulating on the dining room table in West Orange had already reached epic proportions. As the final touches were added a week prior to conferring the exam, the pile reached a critical mass sufficient to force Vasan and his wife to dine each night at restaurants.

Seated at the table each night, Vasan's quest was twofold: First and foremost he had to develop questions pertinent to the course of study being pursued by the Class of 2002 in both lecture and lab. Second, and this is where the textbooks and tracts were of invaluable assistance, he had to find a way to formulate each question succinctly and, most important, without ambiguity.

Vasan needed also to be mindful of striking the balance between the two-hour written exam and the segment of the test most dreaded by the students: a practical portion that required them to answer questions posed on tags attached to structures within forty cadavers. In addition to the tagged bodies, the practical included also ten additional questions affixed to magnetic resonance images (MRIs) illuminated by light tables, human bones resting on auxiliary counters and facsimiles of human organs. Each of the fifty inquiries in the practical were two-part questions, one pertaining to the structure, the second focusing on the attendant clinical function.

Further complicating Vasan's task was the medical school tradition of entering an exam into the public domain. Once administered, the test would be made available to any student wishing to contest a question's veracity, phrasing or relevance. Following the resolution of all challenges, GA exams are set aside for review by subsequent first-year students hoping to divine the future by examining the nature of questions posed in the past. Lest he duplicate a question of recent vintage, Vasan augmented his research with an inspection of anatomy tests dating back three years.

After outlining the shape of the exam, Vasan honed each interrogatory, taking care the syntax was impeccable and double-checking to ensure every structure and function had been brought to the students' attention in either lecture or lab. There was no room for ambiguity, for the test was Vasan's as well.

By stipulating that specifically worded questions be asked on each departmental test, the National Board of Medical Examiners provided the yardstick by which the students, the school and Vasan himself would be evaluated. After the department subjected them to three exams, at the end of the semester the students would face one more test, a National Board of Examiners' shelf exam that served to compare their anatomical aptitude with that of peers across the country. Starting in 1994, when his students averaged seven points below the mean of seventy-eight, Vasan, along with department chairman Dr.

John Siegel, managed to raise the school's score until, in 1998, NJMS edged one point above the national average.

Pleased by the progress, Vasan was far from satisfied. Unwilling to settle for an average score, the course instructor deliberated over the results of each shelf exam, searching for the elements that prevented NJMS from moving to the next level.

The problem didn't reside in recognizing configuration—the students at Newark demonstrated a firm grasp of the structures. Rather, the dilemma lay in a deficiency in comprehending the matching components in an anatomical triptych that included embryology and the clinical correlatives. An appreciation of the anatomical layout was by all means the first and most important step. But it meant nothing without a comprehension of embryological development and, past that, the functional qualities of matured structures. To dwell on one aspect of the course to the exclusion of the other two was to operate in an anatomical void.

"The focus has to be on problem-solving," Vasan explained. "That's why ninety percent of what we teach has to be clinical. They have to be able to think on their feet, they need to focus on specifics. There can be twenty-five reasons why a patient is having difficulty swallowing. We need to teach the anatomy, to complement the physiology, to pinpoint the problem."

The pressure felt by Vasan as he deliberated over the exam, often until one in the morning or later, came from the realization that the exam needed to meet not only the standards of the Board of Medical Examiners but also his colleagues: In the sphere of academic politics, nothing subjected the course director to more criticism and second-guessing than an examination. Some of the faculty, Vasan knew, would find the exam too difficult, others too easy. Always there were rumblings that the students were getting a free pass, that their performance should be evaluated on the basis of more than three departmental exams and that those exams, uniformly, had to be more stringent.

Unmindful of the course coordinator's nightly communion with the examination many believed would shape their destiny, the students certainly didn't pity Vasan the burden of his power over their futures.

As the first of the gross anatomy's three units drew to a close, the dissection of the hand became incidental. In lab, at lunch and in the corridor, the physiology and GA exams were all anyone spoke of. And though everyone dutifully showed up for lab each afternoon, few seemed capable of retaining even a modicum of information.

"Don't ask me about anything below the elbow," Ivan snapped when Jen tried to quiz him on the blood vessels snaking through the wrist. "I'll worry about that after Thursday. My life, right now, has to be physiology."

Ivan had good reason for concern about the second physiology exam. Having come "not even close" to the mean average on the first exam, seventy-five, for Ivan the second exam had been elevated into a make-or-break proposition. The stress showed in his eyes and in the way he pulled away from his lab partners, preferring to peek over the shoulders of friends at other tables rather than to stay put at Table 26.

Sherry thought the engrossment with test scores counterproductive. In the first semester she'd come to the conclusion that, in the manner of most academic tests, med school exams forced students to concentrate on what they guessed the instructors might want them to learn to the exclusion of topics requiring personal development. It was cat and mouse, a game of intrigue that sacrificed enlightenment. "Is this going to make us better doctors?" Sherry mused rhetorically, allowing the question to drift away until the time, if ever the day arrived, when she could take a few moments to consider seriously the misshapen priorities imposed upon her by the medical school.

"I'll be so glad when next Tuesday gets here and this is all over," said Jen, lamenting that three weeks had passed, a lifetime for a twenty-three-year-old, since she'd last spent a night on the town or, in fact, indulged in any activity other than studying. Once the weekend from hell came to an end, Jen vowed to begin a self-reclamation project to upgrade the state of her social life.

Paradoxically, Sherry needed further enhancement of her life outside school like she needed another test that weekend. On the night prior to the physiology test, Jason had a wrestling match. Attendance, as usual, would be mandatory. As was her presence Saturday night at the side of husband, Jerome, at the annual volunteer fire department dinner. Under normal circumstances, Sherry could barely tolerate an affair that transformed her into a captive audience for mothers intent on reciting the minute details of their children's lives. "I'm not trying to be a snob," she explained, "but that's just the way it is. A lot of chitchat and nothing more. Unfortunately, I don't see my kids enough to know how they're doing."

Dreading the dinner more than usual, Sherry doubted her ability to feign even the slightest interest in the attendant conversations as her mind swirled around matters such as the drainage of the intercostal veins, the medial rotation of the anterior fibers of the deltoid muscle and cervical nerve root compressed by the herniation of the intervertebral disc between the fifth and sixth cervical vertebrae. Still, the event meant a lot to Jerome, who, as a cop and a volunteer, interacted with the other members of the fire department both professionally and socially. Given all the sacrifices Jerome had made on her behalf, Sherry knew that attending the function, even coming as it did most inopportunely, was the least she could do.

In any event, complaining only consumed time better dedicated to memorizing a structure that might otherwise have escaped Sherry's attention. Despite her criticism of a system that overemphasized test scores, Sherry couldn't deny that by stressing the importance of review sessions with Vasan and structuring study groups outside of lab, the department was doing its best to prepare the class for what lay ahead.

Inside lab, the preexam preparation one afternoon brought a visitor to Labs A, B, C and D. Such was the distraction level over the test that, initially, no one in Lab C noticed the indistinguishable towel-shrouded lump on the table at the front of the room. Even when they did, from a distance the object wasn't automatically identifiable, notwithstanding the tantalizing hint provided by the tuft of hair sticking out from beneath the fabric.

"Oh . . . my . . . God. Is . . . that . . . what . . . I . . . think . . . it . . . is?" From the far end of the lab, where she had just retrieved a pair of latex gloves from her locker, Leslie Pooser's eyes locked on the table. Steadily, but not too quickly, Leslie strolled to the front of the lab, pulling up to the table at the moment a classmate plucked the towel into the air. For a moment, the pair just stood there, in awe, unable to speak.

"Whoa, I thought I'd seen everything," said Leslie, exhaling loudly. A crowd assembled, their eyes equally wide at the sight of a body prosected laterally at the waist and vertically from the head through the genitalia. At the entrance to Lab D was the body's matching half. The matching halves of another body rested at the front of Labs A and B.

The prosection—the work of David Abkin, whose official title also designated him a prosectionist—was a thing of beauty, an object of revulsion, mesmerizing and horrifying, anatomy in the extreme. The impact was reflected on the students' faces as they inventoried the mirror image of themselves should they ever, heaven forbid, be sliced cleanly in two.

"Like a magician's trick gone bad," someone murmured.

"A nightmare in the making," Leslie agreed. "I *have* to tell my son about this."

But she neglected to tell Omari that night and on all the nights that followed, a disregard unrelated to fear of inspiring a nightmare but the simple fact that, within minutes of seeing the prosection, Leslie completely erased the image from both memory and subconscious: The exams had adumbrated everything, including Leslie's propensity for shock.

As the hours moved steadily toward Friday's physiology exam, time increasingly became of the essence. Each morning Sherry arrived at school with bags under her eyes, tangible evidence of staying up until 1 A.M., two hours past her normal time, while still rising at 5 A.M. Concluding she could absorb more from textbooks than the drone of an instructor, Jen redirected the hours spent in the lecture hall to studying.

Besides, Jen reasoned, there was little need to attend lectures when each day lab was so enlightening, not to mention entertaining, in its own peculiar way. One afternoon, Vasan solemnly announced that someone had been coming into the lab at night and dissecting bodies other than the one assigned him or her. Consider this a warning, Vasan said, that should the practice continue the perpetrator would face dire consequences.

"Why the hell would anyone want to dissect someone else's cadaver? Isn't it bad enough that we have to dissect our own?" Leslie asked, incredulous.

Jen just shook her head. "Beats me; this place is getting weird, very weird."

Right on cue, Roger Faison showed up to reinforce that notion, announcing, with clipboard in hand, that he'd come to "take attendance." The night before someone had dumped an embalmed body with autopsy marks in an alley in the Fourth Ward; unsure as to its origin, the Newark cops contacted the NJMS mortician, who, after inventorying the lab's nonliving population, happily reported back to the sergeant in charge that all cadavers were present and accounted for.

As the exam drew near, Jen's focus intensified. Self-disciplined, she took a no-nonsense approach to studying, checking and double-checking textbooks and notes, relying occasionally on mnemonics, using the first letters of words in an unrelated sentence to enhance memorization.

"Try this one," David Murphy suggested, coming to Jen's aid after she acknowledged a mental block over structures in the wrist: "Some Lovers Try Positions That They Can't Handle." Jen repeated the mnemonic, reciting, simultaneously, the corresponding structures: "Some, scaphoid; Lovers, lunate; Try, triquetrum; Positions, pisiform; That, trapezium; They, trapezoid; Can't, capitate; Handle, hamate." To Murph's amusement Jen, as usual, mispronounced over half the names.

Jen could be excused for butchering the jargon of gross anatomy, an argot that subjected her to terms from a dead language (Latin) and an ancient language (Greek) communicated to her in English by lab instructors with Indian, Hungarian, British, Italian and Brooklyn accents. Cognizant of his thick Indian accent, Vasan enunciated in the extreme whenever he discussed the finer points of anatomy. The course instructor, unfortunately, was an exception, as any student left baffled by lessons conveyed by other heavily accented instructors could attest.

Recounting the mnemonic to David Murphy that afternoon as she waited for the rest of the table to report to lab for a dress rehearsal of Monday's practical, Jen was more inarticulate than normal, a condition she attributed to mental and physical exhaustion rendered by the just-completed physiology exam.

With one test down and the tougher of the two still ahead, Sherry detoured en route to the "practice practical," decompressing at a raucous indoor flea market set up in the school's main hallway. After an onerous morning spent hunched over the physiology test, Sherry welcomed the opportunity to engage, however briefly, in a mindless activity like shopping. Among vendors peddling CDs, Afridesia, perfume, incense and discounted clothing she found two pairs of jeans for the twins.

Enticed by the flea market—the less money she had the more she was tempted to spend it on items she didn't need—Jen bypassed it, heading instead to the lab in order to correspond structures inside the cadaver with notes jotted down the previous night during a fleeting respite from physiology into *Snell's Clinical Anatomy*. "I still feel like I don't know anything," she muttered, forceps squeezing a nerve in one hand and the notes in the other.

"I know what you mean. This is killing me. I can't even remember what we did yesterday," said Sherry, none too pleased, despite her just-completed shopping spree, that the department had mandated her presence in lab ninety minutes after a test on a Friday, normally a free afternoon.

Sherry's frustration resonated throughout the lab, as preternatural medical student carping reached new proportions.

"Why are they doing this to us?" someone moaned, rubbing the exhaustion from his eyes, equating the imposition of consecutive exams with cruel and unusual punishment.

"Because they want to see if we can take it," his friend said resolutely.

If the disgruntled expected commiseration from the instructors, they were looking in the wrong place. For they, too, had been working overtime, readying the lab for the practice practical while the students were sweating the details of physiology, attaching sample questions to specific structures inside the cadavers with string and tags of the type commonly found at garage sales and consignment shops.

The tag attached to a blood vessel in the chest of Number 3426 read: "(A) Identify this structure and (B) It connects to what two structures?"

Ivan scanned the question and announced, "Ligamentum arteriosum. It connects the pulmonary artery and the aorta." Jen, pleased, gave him a thumbs-up.

The practice session required the lab partners to work as a team, moving from table to table within each room, examining the tagged structures and formulating the answers by consensus. Compared with the scene expected during the real thing on Monday, the atmosphere sometimes lapsed into frivolity.

"Where does the vein drain?" Jen asked, reading aloud the question at Table 29.

"In Spain?" Sherry ventured.

"You've been studying too long," Jen said dryly.

The five minutes spent at each table in the lab afforded Ivan, Sherry, Udele and Jen their first opportunity to size up the other cadavers. The more they saw, the more they liked Number 3426.

"Look at these little dinky muscles," Jen remarked scornfully, picking through the minuscule tissue in the upper arm of one of the smaller bodies. "You know, we're going to have a problem with this because we're used to our big muscles."

Circulating through Lab C, a sense of foreboding enveloped the team from Table 26 as they began to understand just how much they didn't know. Ivan, with a two-day growth of beard—"Who has time to shave?"—concluded that, short of sleeping, his entire weekend would be spent in the lab. Had Ivan for some reason chosen to camp out in lab, he certainly wouldn't have been the first, said a faculty member recalling the morning, hours before a practical exam, when he stumbled upon two students fast asleep on the floor next to an open dissection table.

Spurning Vasan's suggestion that the class form small study groups over the weekend, Jen opted again for solitude and announced her only visit to the lab would be Saturday afternoon. She hoped it would be enough. "It's weird; two days before the test and they're asking questions about stuff I've never heard of," she said, aligning herself with Sherry's position about exams being a grand conspiracy lacking relevance to their future abilities as physicians.

The practice exercise did nothing to diminish Sherry's distaste for the examination process. "I might as well quit right now, turn in my papers and forget the whole thing, go back to nursing," she said testily, encountering yet another question about which she hadn't a clue. In the midst of turmoil, Udele remained the picture of serenity: With fewer than seventy-two hours remaining until the exam, Udele knew what had to be done just as she knew, come Monday, that she would do it.

The first entry in Nagaswami Vasan's printed itinerary for Monday, February 8, 1999, began:

"7:30—Set up lab."

And ended twelve hours later:

"7:30—MILLER TIME!!!!!!"

To the minute, Vasan arrived at lab Monday morning to find supply room czar David Abkin and instructor Dr. Paolo Varricchio already converting the four laboratories into testing venues. Knowing from daily observation which cadavers met the criteria for getting tagged, they lined half the tables against the walls in each of the rooms while spacing the remainder across the floor in a manner designed to discourage eyes that might wander to a fellow student's answer sheet.

Vasan had asked the remainder of the instructors to report at eight o'clock; most were early. The faculty knew the drill, fanning through the lab, affixing tags with questions written by Vasan. Opening the cabinet at Table 26, Varricchio exhaled an imprecation upon discovering stained copies of a *Grant's Dissector* and a syllabus. "Didn't we tell them to put the books away?" he asked, tossing the material into an unlocked cabinet, slamming the door shut.

The instructor examined the cadaver and then consulted the tags in his hand, searching for the ideal match between question and structure. "Ah," he murmured triumphantly, fastening the tag to an elastic, white filament under Number 3426's right shoulder: "ID this structure. From which cord of the brachial plexus does it originate?"

After Varricchio knotted the string, he placed a sheet of paper marked "30" on the table's bookstand, converting Table 26, for the purposes of the practical, into Station 30. Quietly, he moved across the aisle to repeat the procedure. "Atrociously boring," he grunted under his breath.

In every room of the lab, instructors mimicked Varricchio,

furtively studying the cadavers and the questions. In the quest for the perfect pairing of inquiry and anatomy, they looked every bit the part of parents setting out Santa's presents under the tree on Christmas Eve.

While taking the test, the students would be given one minute to answer the two related questions at each of the stations. Dispersed between the fifty stations were ten rest stops. For the purpose of the test, the class had been split into three groups that rotated between the written exam in the lecture hall and the practical portion in lab. From start to finish, the practical exam took precisely sixty minutes—of that, Roger Faison made sure.

Once, the timing of the exam had depended haphazardly on the imperfect system of arming an instructor with a stopwatch and a buzzer. Worn by tedium, the timekeeper's laggard thumb on the stopwatch inevitably favored the practical's third shift, sometimes allowing them far in excess of the allotted one minute to answer a question, prompting allegations of an unfair advantage.

Directed by the department to restore equity to the process, the solution came to Faison one afternoon when a solitary stroll along Thirty-fourth Street in midtown Manhattan was interrupted by the low-throated bleat of an emergency vehicle. Hit by an instinctive adrenaline rush, Faison flashed back to his navy submarine service aboard the USS *Flying Fish*, recalling how the sounding of the ship's Klaxon horn caused the entire crew to go into high alert. The medical students, Faison concluded on the spot, needed their own Klaxon horn.

With the assistance of an electrician, by the next practical exam Faison had in place a timer set to a bell that sounded, to the second, after precisely one minute of elapsed time.

As the instructors set up the last of the stations, Faison strolled to the front of Lab C and flicked a switch. With the second hand of his watch, he began counting down. Exactly one minute later, the bell sounded.

"I have to test my baby. I have to make sure that every student leaves here with a piece of me. Long after they're gone, they'll remember this bell, their Klaxon horn," he smiled.

When all the stations were in place, Vasan summoned the faculty to Lab A, where, beginning with station one, they methodically moved through the four rooms, examining each question, double-checking to be sure it complemented the tagged structure, taking care that every inquiry had been presented to the students in either lecture or lab.

"This is Vasan's way of doing it, nobody else in the country does it like this," one instructor said as the pilgrimage got under way. Oblivious, Vasan stepped into the role of an impresario, orchestrating an academic concerto six weeks in the making, only to have the music screech to a halt at Station 5, where questions attached to an angiogram posted on a light table asked the students to identify and correlate the anterior descending coronary artery.

Zolton Spolarics argued that the inquiry—whether the artery extended through the front (anterior) or back (posterior) of the heart's anatomy—was ambiguous. Testily, Vasan countered that the process of reviewing each question and answer in at least four textbooks precluded ambiguity and maintained that the question should remain in the exam.

Huddled around the light table, in the erupting debate Vasan took the position that students needed to be pushed to the limit of analytical comprehension. Adopting the opposing view, Spolarics pointed out, "We're expecting too much here."

For five minutes it continued, with neither side budging until, finally—there were, after all, forty-five questions left to be scrutinized—Vasan capitulated. The inquiry was not too difficult, he said by way of surrender, with the caveat that the question was problematic only in that it posed significant confusion for first-year students inexperienced in the comprehension of radiology charts.

As the group moved to Station 6, Varricchio stole in behind them, switched off the light table, ripped the picture of the coronary artery off the screen and put in its place an MRI of a wrist. From his pocket, he produced the tag with the substitute questions.

Shortly before eleven, the faculty finally completed their rounds. While the wording on other tags spurred minor disagreements, the question at Station 5 was the only one replaced. As the instructors congregated around Station 50, Vasan looked at the clock and dashed off to distribute the twenty-eight-page written exam to Group A, which would spend the two hours from eleven o'clock to one o'clock in the lecture hall. Following a half-hour break for lunch, the contingent was scheduled for the practical, lasting from 1:30 to 2:30. Group A included Ivan and Jen.

Entering the lecture hall, Vasan ignored the intimidating stack of exams and moved purposefully to the front of the room, where he summarily erased review notes inadvertently left on the blackboard. As the course instructor expunged someone's rendition of the thoracic cavity nerve schematic, an anonymous voice from the back of the room yelled out "Honor Code," prompting the last swatch of levity Group A would know until mid-afternoon.

Assured that everyone had an exam, Vasan started to make his way back to the lab, only to be met at the door by a wild-eyed, rumpled and disheveled student arriving five minutes late for what he knew to be the biggest exam of his life: Ivan.

Forty minutes later, a full twenty minutes before the first shift of the practical was scheduled to begin, the students assigned to Group C began gathering outside the lab, filling the corridor with anxious chatter. A short time later, Udele joined them, her manner reflecting none of the nervous tension endemic among the others. At 11:55, the lab doors swung open; as they passed through, each student was handed a clipboard with a page numbered 1 through 50.

"Find a chair, any chair, write your name at the top of the paper and then write nothing else," instructor Anthony Boccabella commanded.

Once everyone was in place, the public address system crackled to life. "Be sure to start answering the questions at the number of the station where you begin," Varricchio warned. "Otherwise, you will run into major problems when we begin to grade. Move quickly between the bells. Don't talk among yourselves. Good luck."

"And for heaven's sake, don't pull on the strings," Boccabella added.

A few smiled at the instructor's humorous aside; most didn't even hear him. The students seated at stations with questions contemplated the tags in front of them. The students at rest stops stared vacantly at the wall. But for an occasional sigh of released tension, the room was completely still, a collective concentration so keen that no one even noticed Faison as he moved to the wall at the end of Lab C, where, after receiving a discreet nod from Boccabella, he threw the switch that, ten seconds later, set in motion the intellectual calisthenic that had weighed on the class from the onset.

CHAPTER 11

———————•———————

As the migraine spread from his frontal sinus back to the ears, Ivan Gonzalez squeezed his eyes shut, willing the lecture hall to stop spinning, silently cursing his luck, trying to imagine a more inopportune moment to be trapped in a vise of pain.

Had his medical training ended at that very minute—and the harder he fought to concentrate on the gross anatomy written exam the more he feared it might—Ivan still could have managed a self-diagnosis. The origin of the headache, as doctors had explained to him at the onset of migraines while he was in college, was stress, triggered in this instance by Ivan's conviction that his entire future depended on the outcome of the test.

The stakes had been raised midway through the physiology exam on Friday with the realization he was doing no better on the second physiology test than he'd done on the first. Convinced he was on the cusp of a no return in one class, Ivan compensated by engaging in a frenzied struggle to prevent similar disaster in the other, gross anatomy.

Despite a weekend spent in a nonstop volley among the lab, study groups and the empty classroom where he devoted hours to memorizing mnemonics and crude anatomical schematics scribbled on the erasable wallboard, Ivan still managed to squeeze in a decent amount of sleep on Friday, Saturday and even Sunday night. The balance between studying and proper rest seemed for naught when midway through the written exam his head began to throb so severely that Ivan wondered if he would be able to finish. Somehow he managed, and at 1 P.M., Ivan staggered from the lecture hall, his eyes red slits, looking every bit a drunk coming off a four-day bender.

"My brain is aching," he whispered huskily, collapsing into a chair, knowing the half-hour before he had to report for the practical exam offered far from enough time to recoup.

Emotionally and physically depleted, twenty minutes later Ivan fell into place outside the lab, biding his time until the doors swung open, allowing him entry to a test that, if the pattern continued, would stop dead in its tracks a dream that had forever flown in the face of reason.

Compared with the Nicaraguan norm, the family of Francisco and Anna Maria Gonzalez was not far separated from the lap of luxury. Amply supplemented by the income Anna Maria brought in as a clothing merchant, Francisco's salary as a route supervisor for Pepsi-Cola allowed for many amenities unavailable to friends, neighbors or, for that matter, the majority of their countrymen.

Home was a two-bedroom, S-shaped house on the outskirts of Managua surrounded by mango trees; Francisco and Anna Maria occupied the master bedroom, their seven children—four boys and three girls—shared the other, two to a bed.

Most Nicaraguans couldn't afford one car; the Gonzalezes had two, used for mandatory transportation as well as vacations to the mountains, nearby lakes and the Pacific Coast, two hours from their home, a residence enhanced by both a television set *and* a stereo system. For her children's birthdays, Anna Marie purchased cakes from the local bakery and decorated the house with balloons and streamers. Commonplace celebratory icons in the United States, the embellishments were considered an extravagance in Nicaragua.

Living in a nation rife with poverty, the Gonzalezes fell squarely in the economic class known colloquially as "Borgeses"—middle class, bourgeois, a status many of their countrymen found contemptible.

Born in 1974, Ivan was the youngest and the most sickly, plagued until age eleven by asthma and bronchitis, medical conditions that lumped yet another layer of concern onto a family already buffeted by the political unrest seeping into so many other Nicaraguan households.

In a neighborhood polarized by an unfolding civil war, Francisco and Anna Maria Gonzalez remained steadfast pacifists, sympathetic to neither the regime of despotic president Anastasia Somoza or to the Sandinistas, the Marxist insurgents mounting an escalating campaign of violence against the Somozan government. Serving the family well as an ideological concept, the practical application of pacifism became hopelessly ineffectual once the conflict moved out of the mountains to land, quite literally, on the Gonzalezes' front door.

Unable to recruit numbers sufficient to guarantee Somoza's downfall, the Sandinistas adopted a campaign of conscription that entailed kidnapping able-bodied young men, and sometimes young women, off the street at gunpoint, forcing them blindfolded into military vehicles that spirited them to the mountain hideaways that served as indoctrination camps.

Basic training, Sandinista-style, often took upward of two years. Male conscripts spent days digging ditches and nights becoming indoctrinated in Marxist ideology; women recruits harvested coffee beans along with Marxist dogma. Regardless of whether the inculcation took hold, it paid to at least feign allegiance to the cause. Among the draftees, it was well understood that those who played along were allowed visits from their families, while those who didn't often disappeared forever.

The Sandinistas' reign of terror was by no means limited to the streets. At least once a day, and sometimes more, guerrilla troops embarked on a neighborhood census, demanding that every occupant of every residence be accounted for. When the soldiers came calling, young men residing within the environs of each house understood intrinsically that betrayal of their presence meant a fast trip to the nearest Sandinista encampment.

At the Gonzalez home, the troops had to first pass through Anna Maria, who stood sentry at the front door, denying access to the interior of what had become a mango-shaded bunker protecting her three eldest sons. Placed under house arrest by their own mother, Francisco Jr.—"Frankie"—Omar Antonio and Roberto Manuel chafed at imprisonment. The conscription of their sister, Lenora Del Socra, only heightened their mother's resolve. And when Omar voluntarily joined ranks with the Sandinistas, hoping his own sacrifice would allow his brother Frankie to attend college, Anna Maria—the matriarch in every sense— knew what needed to be done.

Before their own country fractured, the United States held no allure for Francisco and Anna Maria Gonzalez, who, through ambition and hard work, had managed to fashion the North American ideal on Nicaraguan soil. The two-day treks to visit Omar in the mountains, and not knowing if Frankie or Roberto would be there when she returned home, convinced Anna Maria of a truth long avoided. To survive, the Gonzalezes had no choice but to flee, one by one, until all nine family members were Nicaraguans only in heritage.

Reasoning that "It's terrible to leave your country, but it is worse to [have your children] leave [permanently]," the decision to migrate was made with the inherent knowledge that the odds of the entire family reuniting in North America were at best minimal. Like many Nicaraguans, the Gonzalezes could name hundreds of countrymen who'd escaped to the United States and also a fair share who had perished trying.

Frankie, the most vulnerable son, was the first to go, slipping in 1984 through the Nicaraguan countryside to Guatemala and then Honduras, where he boarded the excursion bus renowned for issuing round-trip tickets that invariably went half-used. "No one who ever left on that bus came back on that bus," Ivan recalled.

On its journey through the Guatemalan countryside the bus was regularly halted by armed soldiers demanding that the passengers surrender their passports. Twenty U.S. dollars placed in a coffee can passed from seat to seat was the going price for those wishing to have their passport returned.

When the bus dipped into Mexico its occupants disembarked under the guise of seeing the sights and headed instead across the country, bound for the Rio Grande and, beyond that, their final destination, Texas. Upon Frankie's departure his family was unaware that the trip from Managua to the northernmost Mexican border took anywhere from six to nine weeks. As the weeks dragged into months with no word from her oldest son, Anna Maria began to assume the worst.

Death, it turned out, was about all Frankie avoided during his passage into the United States. Robbed of all his money, he arrived in Brownsville, Texas, with not even the shirt on his back—that, too, had been stolen. In Brownsville, he convinced a newly made acquaintance to lend him a few dollars against future earnings—how, precisely, those wages would be secured he hadn't the slightest idea. Still soaking wet, he used the money to call home.

Anna Maria answered the telephone and heard a voice quivering so profoundly she barely recognized it. "I'm alive. I can't talk, Mom, I've got to go"—that was all Frankie could afford to say before the operator disconnected the call. Anna Maria stared at the receiver and sank to her knees, sobbing uncontrollably.

One by one, the others left: Maria Elena, the oldest; Lenora Del Socra; Roberto. In May 1985 came Ivan's turn. Eleven years old and unable to make the odyssey on his own, he was entrusted into the care of his brother-in-law, Felipe Jarquin, Maria Elena's husband.

Following Frankie's route to Texas, Ivan and Felipe encountered identical problems, with Guatemalan soldiers and other human parasites preying on the vulnerability of refugees in flight. When finally they arrived in Mexico, the boy and his protector waited well into the morning at a prearranged site for the coyote, the Mexican border guide, who'd promised, for a fee of a hundred dollars American, to facilitate their safe but illegal passage into the United States. Their light clothing provided only the barest protection from the nighttime cold; when the sun came up it become clear the rendezvous would never occur. Wearily, Felipe towed his charge to a hotel where he spent $20 on a room they could ill afford to get the sleep necessary to ensure an alert and safe passage as soon as darkness again descended.

That night, Felipe and Ivan returned to the Rio Grande, where they located a maze of drainage aqueducts spanning the river. Stripping bare—wet clothes were a red flag to U.S. Immigration and Naturalization Service patrols—Ivan climbed atop Felipe's shoulders and, in the dank of the tunnel, with water rushing around them, they began making their way across.

When the water rose to the level of his brother-in-law's chest, Ivan wasn't sure they would make it. Somehow they did. Once ashore in Texas they scrambled to get dressed and made their way into Brownsville. After contacting Maria Elena, who was then in Miami, the pair moved along to the Brownsville bus station, where Felipe purchased two one-way tickets to Florida. Following the transaction, three dollars remained in his pocket.

As they waited to depart, an immigration officer climbed aboard the bus, requesting in Spanish that the passengers turn over their passports for inspection. When he reached the Nicaraguans, Felipe nodded toward Ivan, slumped in an exhausted sleep. The passports and immigration papers, Felipe indicated, were in the possession of the boy. The officer studied the slumbering child and moved as though to wake him, halting at the last moment to inquire their destination. Provided the information, the officer hesitated for what seemed an eternity, before curtly ordering Felipe to "take care" of the matter.

It was not by design that the Gonzalez family wound up in South Florida. Arriving in Brownsville, Frankie and a handful of other refugees newly emerged from the river were hired, on the cheap, by a local contractor to complete a roofing job. Satisfied with their work, the contractor brought the crew on board until such time as they'd raised enough money to leave Texas. When that day arrived, Frankie's fellow travelers set out for their original destination, Miami. Knowing not a single other soul in the United States, Frankie joined them.

A year later, the immigration official placated, the bus pulled out of Brownsville, taking Ivan and Felipe on the final leg of their exodus. In Arkansas, still two days from their destination, Felipe spent the last of the money on sodas and crackers. During their next layover, in Alabama, a service station clerk noticed the bedraggled eleven-year-old making his way to the rest room and thrust a handful of candy bars at him as he made his way back to the bus. When Ivan threw his hands in the air, indicating he had no money, the clerk smiled, handed him two more candy bars, and shooed him on his way.

Slowly, the rest of the family trickled into Florida. Francisco and Anna Maria waited until all the children were gone before departing themselves. In December 1986, they walked out of the home where both believed they would live forever, leaving behind everything but two albums stuffed with family photographs. Before she headed for a country whose materialism she feared would destroy every value instilled in her children, Anna Maria painstakingly sewed American dollars into the hems of the three layers of clothing she wore on the pilgrimage.

Avoiding the arduous path taken by their children, Ivan's parents flew circuitously from Costa Rica to Mexico. Still, nearly a month passed before they reached the Rio Grande, where, just feet from the river's edge, they were intercepted by the Mexican border patrol. Five hundred dollars later, the journey resumed with Francisco, then fifty-nine, and Anna Maria, forty-five, jumping from boulder to boulder in an effort to avoid the river's rushing current. The night was cold, the crossing treacherous. Halfway across, Anna Maria lost her footing and tumbled into the water; instinctively another of the refugees reached in and latched on to the nearest part of her body: her hair. Fighting the river, Anna Maria's rescuer tugged until he was able to get an arm around her shoulder and pull her to safety.

Two days later, the Gonzalez family—father, mother, seven children and a couple new grandchildren as well—were reunited in a Miami neighborhood that preferred not to have them. No one's idea of paradise even in its heyday as a mecca for Cuban expatriates, by the time the Gonzalezes turned up, Little Havana was in a precipitous decline. Taking the place of upwardly mobile Cubans transplanted to suburban Dade County, poor urban African Americans' and recent Central American immigrants' migration into the neighborhood had been anything but harmonious. The African Americans resented "wetback" Central Americans for stealing their jobs, the Central Americans despised the African Americans for their intolerance, and the entrenched Cubans begrudged all parties for infringing on their turf.

Too young to help his family establish a financial foothold, Ivan nonetheless faced a task nearly as formidable: Alone among his parents, brothers and sisters, it fell upon him to learn the native language of the new country.

Keeping a promise to her mother that Ivan would be held to a degree of discipline in Miami proportionate to that meted out by Anna Maria in Managua, Maria Elena rode herd on her younger brother, enrolling him in an English as a Second Language course at the neighborhood elementary school within six weeks of his arrival. Ivan walked into the school knowing precisely six words of English: "My name is Ivan" and "Thank you."

For Ivan, becoming proficient in English proved to be onerous, impeded in no small part by his tendency to lapse, outside the classroom, into the only language spoken at home, Spanish. Socially, he fared little better. Ivan always had been, in Anna Maria's words, "flighty," citing the afternoon in Managua when he had been left in the care of an older sister while the rest of the family sneaked out to run errands. Responding to a knock on the door, the sister found herself face-to-face with three armed Sandinistas demanding access to the house. She allowed them entry and, for five harrowing minutes, hovered protectively over the child as rebels went room to room searching for the older Gonzalez brothers. Alarmed by the potential impact of the episode on his psyche, Anna Maria later interrogated her son about the episode. Ivan recalled nothing; riveted by a television show, he hadn't even noticed the intruders.

Preadolescent awkwardness, his emerging intelligence and unfamiliarity with the language conspired with his distracted personality to exacerbate Ivan's adjustment problems as he struggled mightily to fit in at Robert E. Lee Elementary School. Studying became his coping mechanism, and he found solace in science and math, subjects that, not coincidentally, transcended the language barrier.

To classmates who viewed academic excellence as a character flaw, Ivan's acuity became a flash point for interminable harassment that only forced him deeper within himself. The nadir occurred during Ivan's first week at Jackson High School when a bully blackened his eye. True to form, he went home without reporting the incident to school authorities and refused to divulge the details to anyone, even his mother. Appealing to the instinct for self-preservation, Anna Maria finally convinced Ivan to at least share his travails with the school's guidance counselors.

The depth of Ivan's predicament tugged at the counselor, Iris Enteen. Assigned to one of Miami's toughest high schools, Enteen rarely saw the likes of an Ivan Gonzalez, a young man who knew at age fourteen what he wanted to do with his life and possessed the wherewithal to do something about it. Enteen nonetheless had good reason to be wary: Too often she'd seen what became of young people like Ivan in an atmosphere where social rewards, in the form of popularity, were usually meted out to the school's lowest common denominator. Still, the untapped resolve Enteen identified in the painfully shy young man struggling to converse fluently in English persuaded the guidance counselor that it would be worth her while to keep an eye on Ivan during his four years at Jackson.

At their first meeting, Ivan informed Enteen that he'd one day be a physician. It would take a while, but by Ivan's junior year Enteen became certain it would happen. Not that there weren't compromises: Survival dictated that whenever possible Ivan mask his intelligence by acting as if he operated at a level of ignorance equal to that of most of his classmates, a ploy that meant the academic trophies for science and poetry prominently displayed in the Gonzalezes' living room were never mentioned in the halls of Jackson High.

Ivan emerged from high school with a perfect attendance record, recognition from the National Honor Society and a scholarship at Barry University in Miami Shores. Medical school was finally within his sights. Compared with what he'd gone through in order to attend college in the United States, Ivan found the university experience itself relatively easy, sailing through most of his classes, rekindling an affinity for the life sciences on his way to a cumulative 3.4 grade point average. As far as he was concerned, Barry had only one drawback: Twenty years after the all-female school first admitted men, "all the good-looking women were gone."

As graduation from Barry University drew near, Ivan encountered resistance about attending med school from a most unlikely source— his mother. Disconcerted by the lifetime of trauma already visited upon her children, Anna Maria pleaded with Ivan not to pursue a career that would steal another seven years from a life already inter-rupted.

Ivan rejected out of hand Anna Maria's entreaties that he settle for a job in a research lab. Scientific research entailed writing "crap" that, in most cases, made not the slightest bit of difference in the life of anyone. Put bluntly, research was not medicine. Ever his stubborn mother's son, Ivan went ahead and took the Medical College Aptitude Test. While an MCAT score of 25 ruled out admission to the pantheon of top medical schools, it didn't preclude some second-tier institutions from considering a prospective student who might bring to their institutions a most intriguing background. Ivan's Dickensian biogra-phy did him not the slightest bit of good when the first choice on his list, the University of Miami School of Medicine, red-flagged his application due to a circumstance having nothing to do with personal or academic qualifications: Ivan, it turned out, was still an illegal alien.

Rejected for asylum immediately after arriving in Miami, Ivan had become so absorbed in the process of becoming an American in every other sense that he'd neglected to seriously pursue the matter. Now, eleven years later, the negligence fulfilled his mother's wish. Placing medical school on the back burner, Ivan accepted a position with a genetic research firm and began anew the process of becoming a citizen. Nearly a year passed before Ivan's lawyer convinced a federal immigration judge that his client "did not come here to feed off the system." It was March 1997, ten months since Ivan's college graduation and a month beyond the deadline for entering the University of Miami or any other medical school the following fall.

When autumn arrived, Ivan began applying anew and was stung to again be summarily rejected by the University of Miami. Dejected, he took a chance on a long shot and fired off an application to UCLA; sufficiently impressed, the school flew him out for an interview. Following introductory pleasantries, the first question posed to the applicant had to do with the population of Managua and the current status of the Nicaraguan economy. Ivan's failure to deliver an intelligible answer to either question got the interview off to a poor start. And from there it went downhill. "Everybody tries to Americanize you, but then they want to know about your roots. How was I supposed to know? I'd been gone twelve years and I was only a kid when I left," he complained afterward.

Flying back to Miami, Ivan was ready to give up. Applications to other institutions had also drawn letters of rejection, causing him to believe med school "wasn't meant to be." Urged by a friend to give it one last stab, he sent applications to the New York Medical College, the Albert Einstein College of Medicine and, at the last minute, a place he had never heard of, the New Jersey Medical School.

Ivan's appreciation for public health facilities, dating back to the clinic that brought him respite from the bronchitis and asthma, had broadened in the United States after his witnessing the efforts performed by the system on behalf of Anna Maria, stricken with heart problems after arriving in Florida. NJMS's connection to University Hospital thus made the school the latest in a series of first choices, a status that doubled Ivan's elation upon receiving the call from George Heinrich announcing he'd been accepted.

Any joy Ivan felt over finally making it into med school evaporated the moment he set up house in a Newark apartment midway through August 1998. Ubiquitous heat and humidity may have been part of growing up in Central America and Florida, but the apartment's lack of air-conditioning so tested the limit of Ivan's tolerance that he spent much of his first night in New Jersey standing under a cold shower. Loneliness, disorientation and, chiefly, nagging concerns over money only added to his discomfort.

Although geographical proximity to friends and family had played a role in Ivan's predilection toward attending med school in Miami, the prospect of paying in-state tuition, while defraying other expenses by living at home, weighed equally. Compared with Miami, NJMS was a money pit: Ivan's $32,000 annual tuition, coupled with the ancillary expenses, meant his disbursements far exceeded those paid by in-state students. Furthermore, the cost of living imposed by New Jersey's proximity to the money corridors of downtown Manhattan caused Ivan to start out in a financial hole that promised to grow bigger by the month.

Powered by a genial personality, Ivan combated his loneliness during the first term through basketball and teaching rudimentary Spanish to classmates twice a week over lunch. The financial conundrum was not so easily solved. Ivan's brothers and sisters helped out where they could, but because they had families of their own to care for, Ivan loathed turning to them and sought their assistance only when both the refrigerator and the gas tank teetered perilously close to empty.

Exacerbating the ongoing vexation about money was school itself: Two years in the workaday world had left Ivan academically rusty. Thrust into an atmosphere that brooked no alternatives when it came to studying, Ivan found it difficult to balance the demands placed upon him by the school with what, for the previous twenty-four months, had been an abundance of leisure time.

Returning to the classroom, the first semester diet of tissue and cell biology and biochemistry, two of his strong points, proved far from overwhelming. There was still time for basketball and friends, Spanish tutoring and, when he had enough money, after-hour excursions to medical school hangouts. Still on cruise control, an emboldened Ivan entered winter term expecting a few more bumps than he'd encountered in the first semester, but a road nonetheless navigable.

The hard part had been getting into medical school. Having snatched the brass ring, Ivan showed up in January primed for the straight shot to medical residency and beyond. Within a month, he was in the ditch.

Seated in the lecture hall, the agony of the migraine searing as the fear grew that his flame-out in gross anatomy would be even more spectacular than the one he was currently experiencing in physiology, Ivan struggled to focus on the written exam. Questions he could have answered easily twelve hours before were indecipherable; a veil of uncertainty descended, intensifying the pain shooting through his forehead. When after precisely two hours the proctor asked the group to put down their pencils, Ivan felt neither exhilaration nor relief, only numbness. His confidence shot, he staggered toward the lab spiritually bankrupt, certain the practical would be the coda to failure.

Clipboard and answer sheet in hand, Ivan climbed onto the stool at Station 45 and immediately went about getting Jen's attention at Station 43. Ivan, she noticed, looked deflated. Jen was well aware of Ivan's struggles in physiology and, from the looks of things, now also gross anatomy. Jen acknowledged Ivan with a quick smile and returned her attention to the tag on the cadaver. It was not an opportune moment to commiserate over Ivan's woes. Besides, Jen had worries in a similar vein: After never even coming close to flunking a test in seventeen years of schooling, Jen had departed the written exam with a gut feeling her streak had come to an end.

As Jen studied the cadaver at Station 43 she experienced another wave of doubt. The body in front of her seemed foreign, nothing like Number 3426, whose landmarks bore a comforting familiarity.

Initial disorientation in a practical was a phenomenon often seen by veteran instructor Anthony Boccabella, who compared the sensation to occupying the rear passenger seat on a trip to an unfamiliar destination and then taking over as driver for the return trip: The bridges, groves of trees and other landmarks encountered might seem vaguely familiar, but the context has changed dramatically.

As the last of Ivan and Jen's group took their places at the fifty stations, Varricchio repeated word for word the instructions he'd delivered to the first contingent. Then he wished everyone luck and Faison sounded his Klaxon horn.

For the next hour, the horn provided the sole diversion to the singular attention demanded of the forty-five students assigned to Group A. With each sounding, the group shuffled dronelike to the next station, where, with every ounce of concentration riveted to the questions on tagged structures, sixty seconds ticked incessantly away in their minds. The answers either came immediately or in the instant before the horn rang. Rarely did anyone jot anything down after twenty, thirty or forty seconds had elapsed. "If you didn't get it right away, you only had a few seconds to make something up. That's not a lot of time to make something up," Jen later observed.

Ivan arrived at Station 50, a rest stop, six minutes into the practical with his brown eyes alert, vivid, reflecting not the slightest trace of the exhaustion evident when he'd entered the lab. Before he climbed onto the stool Ivan again caught Jen's attention and did a little jig.

While his classmates labored for sixty seconds, Ivan stretched backward and closed his eyes, the one-minute respite interrupted by Boccabella, who leaned over and whispered, "We don't really grade this, you know. It's just a tradition we put you through because we like to watch you sweat and we want to give you something to brag to your friends about after you get out of medical school."

Ivan attempted unsuccessfully to stifle a chuckle. The veil had lifted the instant Ivan took his place at Station 45 and read the tag attached to the cadaver's heart: "Identify this structure." Hesitating not a moment, Ivan knew the answer: "the muscular interventricular septum." He gazed at the second question: "Which part of this structure is the most frequent site of septal defects?" Again, the answer came instantaneously: "membranous."

An identical experience occurred at Station 46, where a capillary attached to the trachea had been tagged. "Identify this vessel." Ivan wrote, "brachiocephalic trunk." "In which mediastinal trunk is it located?" Ivan jotted down, "superior."

Proving true Jen's theory of spontaneous recognition, the automatic responses came again at Station 47, then 48 and 49. At Station 50, the rest stop, Ivan contemplated where he'd been and where he was going. The results of the first two physiology exams as well as the morning's written exam, which he also assumed he'd bombed, could not be minimized; he'd have to redouble his efforts the rest of the semester to overcome the hole he'd dug for himself.

The first five questions in the practical, even if he'd been mistaken, proved again his resilience. A corner had been turned. The horn sounded and Ivan moved away from the rest stop to Station 1, his headache gone, replaced by a confidence he hadn't felt in over two years.

Forty-five minutes later, when it was over, the atmosphere felt like air expended from an oversized balloon.

"Yeah, yeah, yeah," Ivan chortled, striding through the corridor outside lab, an enthusiasm tempered by the sight of Sherry, anxiously waiting for her turn to take the practical.

"Everybody loved our body," Jen whispered to her lab partner. "They thought we did a great dissection. I'd like to take all the credit, but . . ."

Sherry, her eyes only slightly more alert than Ivan's had been an hour previous, barely acknowledged what Jen was saying.

Having survived the practical, Ivan and Jen conferred with their classmates about the precise location of the Hoboken bar that was the site of the postexam festivities. As a concession to requisite afternoon naps after what, for most, had been four relatively sleepless nights, a starting time of 9:30 was established.

A voice of protest questioned the wisdom of drinking copious amounts of alcohol on the eve of a procedure that promised to equal, if not exceed, the emotional distress wrought by the first cut and the flipping of the body. Jen rolled her eyes at the protestation.

"After this dehumanizing day I'm not going to get wigged out about the head tomorrow," she promised, heading home for a nap and her first study-free night in weeks.

CHAPTER 12

———————•———————

The next day, the number of tables that contrived to dissect the neck without removing the towel from the cadaver's face proved Jen as an exception. Indeed, a good deal of the class remained quite wigged out by the head. "I know it sounds trite and I know he's dead, but you know what they say, the eyes are the windows of the soul," explained Annette Pham, who went to great lengths to ensure the towel didn't slip at Table 23. "It's difficult to expose the eyes because the eyes make it more real."

A similar concern was not in the least evident at Table 26, where, with the face exposed, not once did anyone mention that their every move fell under the gaze of eyes shielded by lids half shut, a nonchalance attributable in part to the cadaver's incidental presence. For on an afternoon when due attention was supposed to be paid to the posterior triangle of the neck, Table 26 instead joined the rest of the Class of 2002 in dwelling on the grueling circumstances imposed upon them over the weekend.

Analogies about the punishing physiology-anatomy exam tandem ranged from fraternity hazing to Marine Corps boot camp. Christine Ortiz swore she'd exhibited symptoms of post-traumatic stress disorder in the aftermath of the practical. Alex Flaxman said the exams "Ate me up, spit me out, tossed me in the trash and threw me in the alley where a dog peed on me."

Exasperated by the grousing, Jen did her best to put the exams into a context slightly less dramatic than Armageddon: "I know people are upset, but what's the big deal? It's *supposed* to be hard. If it were easy, then everybody would do it. I don't know what they expected this was going to be like."

Amenable to Jen's morsel of perspective, Sherry still called it "the hardest test I ever took" with the caveat, "Then again, I always say that." She'd ducked out of the fire department banquet early Saturday night, returning home to study until the wee hours of the morning. Before the kids were out of bed, Sherry made the trek to Newark, where she spent the entire day closeted in a private study cubicle or in the lab. The effort, she feared, hadn't been enough: "I screwed the whole thing up. I wasn't ready."

"But today we get to start all over," Jen said. Despite the feeling she'd done exceedingly well in the practical segment of the exam, Jen still feared the outcome of her overall grade.

Uncharacteristically letting down her guard, Udele lamented the psychological gamesmanship that had caused her grade to plummet. Taking a reactive approach, at each station in the practical Udele tended to respond within ten seconds of encountering the question, a technique that resulted in her playing a mind game until the horn sounded. Twice, Udele second-guessed herself during the remaining seconds before changing her responses at the last possible moment. Following the practical, she immediately consulted her notes. Both times, the answer born of instinct had been the correct one.

Nothing could be done to right the wrong except to continue moving forward. Udele sighed and began to flip absently through the syllabus. Ignoring the ongoing postexam retrospective at Table 26, now joined in progress by Christine Ortiz, she made three perfunctory incisions in the neck before indifferently tugging the latex gloves off her hands. The unspoken announcement that Udele was of neither mind nor mood to continue stopped the other conversation cold. Udele walking away from the opportunity to dissect was a development of major proportions. "We couldn't do this today either," Christine reassured her and, with that, Table 26 joined the rest of the class in calling it a day.

As she departed lab, Jen sidestepped the throng at the corridor bulletin board straining to see the grades for the written portion of the gross anatomy exam. Making a beeline for her locker, she made a mental note to get to school early the next morning, preferring to be alone when confronting what promised to be an unwelcome reality.

"It's just a shave. I keep telling myself it's just a shave, a little deep but it's still just a shave," said Leslie Pooser. By blowing off lab Tuesday, the class set itself up for a double workload on Wednesday. Leslie, true to form, inaugurated the onset of the newest procedure by articulating her trepidation.

Instructor Anthony Boccabella found the ritualistic angst each year over the onset of Unit II—four solid weeks dedicated solely to the examination of the head and neck—a superfluous distraction. "A lot of emotional energy is expended on the face and neck, but in the long run, it's the cavities and extremities that command the most attention once they enter the field," said Boccabella. In another week, when the skin of the face had literally been peeled from the bone, the instructor predicted clinical interpretation would again supersede sentiment.

At ever-pragmatic Table 26, the initial stage of the second unit presented problems not of sentimentality but of logistics. Unlike the broad expanses of Number 3426's chest, arms and back that had allowed two lab partners to labor simultaneously, the neck provided a much smaller canvas upon which to work, a situation that promised greatly to curtail individual cutting time. As would be the case throughout Unit II, the neck also required an exactitude missing from the first unit. In other words, the area left little room for error.

"In the neck, each centimeter is a new world," Vasan explained to them. "That is why the manifestation of injuries found in the neck are so varied: tumors, traumatic injuries—all of them have a different impact."

Trauma to the neck was a topic familiar to Sherry and, upon the reflection of the skin, she offered a brief dissertation on the "blind" tracheotomy. Rummaging through the thicket of muscles, tendons, nerves and capillaries, she demonstrated how the successful insertion of a breathing tube during emergencies is more often a product of luck than skill.

The presentation complete, Sherry retracted the sternohyoid muscle to reveal the laryngeal prominence. Known in layman's parlance as the Adam's apple, its presence in the male cadaver's neck sent Table 26 scampering to Table 27. As they suspected, though the laryngeal cartilage in the female cadaver was detectable, it could not be described as prominent.

In the ensuing conversation about the protrusion that betrays the gender of even the most fastidious transvestite, no one noticed that Nagaswami Vasan had slipped quietly to the front of the lab. With him was Paolo Varricchio, wheeling a cart nearly overflowing with stacks of papers: the graded practical exams. One by one, each table spotted the course coordinator, causing an uncomfortable silence to ripple through the room. Having gained their attention without employing the traditional hand clap, Vasan launched into a pep talk:

"This was a very powerful exam. Very powerful. It stands up to any exam given at any medical school in the country. Ninety percent of the questions concentrated on problem solving and the results showed that you are ready to solve problems. Your weak spot is embryology. You have to start attending lectures again. You're blowing off embryology; that's like blowing off eighty percent of the exam."

Jen looked at Ivan and shrugged.

Vasan continued: "The average was seventy-seven/seventy-eight. That's good, but not good enough. The mandatory attendance in lab helped; it shows you need to be here, that you need to get the experience and to learn from the faculty and from your lab partners. You can't do this alone. I know you had to study at the last minute; I know about the physiology [exam] right on top of this one. That won't happen again.

"OK, there's three weeks before the next exam. Three weeks. Between now and then, we're going to cover two hundred pages in *Snell.* Use every minute you're in lab to discuss anatomy. Don't use this time to discuss basketball. You can only talk about anatomy when you're in this room."

"Oh shit, we can't discuss the Knicks?" someone muttered at the back of the room.

Vasan didn't hear. "You are entering the most difficult part of the anatomy. There is limited space in the neck and head, but stay close while your lab partners dissect, pay attention, stay the course, don't leave every day at four and don't wait until the last day to study."

The coordinator handed the graded written and practical exams to Varricchio for distribution and, turning on his heel with the resoluteness of a commanding military officer, marched off to deliver the same spiel to the troops in Lab D.

Privately, Vasan was more than satisfied with the mean score of 78, a performance seven points higher than the average on the national anatomy shelf exam and ten points higher than the scores recorded by NJMS students ten years before. Reflecting grades neither preponderantly higher nor lower than the average, Vasan felt the mean proved the exam's inherent fairness. Still, unwilling to stand pat, Vasan pulled out the exam the next day, looking for ways in the future to raise the average to 80.

Not terribly eagerly, Lab C shuffled over to Varricchio to learn their fate. Wishing for the opportunity to ascertain her performance in private, Jen waited until everyone else had received their scores before retrieving her exam. Purposefully taking her time returning to Table 26, she peeked at the numeric figure written at the bottom of the practical, automatically adding it to the score she'd seen that morning on the bulletin board. The number caused her to halt in her tracks, her face turning crimson.

"You OK?" someone asked.

"Yeah . . . fine."

Recovering her composure Jen returned to Table 26, where everyone, save Ivan, had donned masks of nondisclosure. To look at her, no one could have told that Udele had scored on the cusp of the class elite. Or that Sherry had finished well above the mean, although not as high as she'd hoped—"I've sort of given up on getting honors," she admitted later.

Again violating the unwritten rule about openly discussing grades, Ivan let slip that a horrendous showing on the written test notwithstanding, he'd managed to parlay his performance on the practical into a grade just two points below the average score. Ivan had dodged the bullet, but not by much: Before departing school later in the afternoon he discovered in his mailbox a missive from Vasan, summoning him to a series of special early-morning tutoring sessions to help him prepare for the next exam.

Despite portents of failure, Jen more than dodged the bullet; it missed her by a mile. Convinced she'd flunked a test for the first time in her life, Jen instead had passed with high honors. Reading the score of the written exam on the bulletin board that morning, Jen's disbelief had been such that she'd used a sheet of paper to confirm that the grade and her Social Security number shared the same line. Pride of accomplishment, however, came with a price Jen could ill afford: Having lost a bet with a classmate that she would finish with the lower grade, Jen worried that her checking account contained insufficient funds to cover the ante, dinner for two at a restaurant of the winner's, so to speak, choice.

To the advantage of one party and the disadvantage of the three others at Table 26, medical school is a venue providing not a nanosecond for its apprentices to exult over achievement or wallow in failure. Meaning by Thursday the exam and resulting grades had dropped entirely from conversation topics, butted aside by the realization that, before afternoon's end, everyone in the class would play a role in erasing the face from a human being.

"If you never dreamed about it before, you're going to dream about it now," Leslie predicted.

A studied calm came over Table 26 as Udele picked up the scalpel and followed, step by step, the directions dictated by the *Grant's* lying open on the towel covering Number 3426's chest: "In the midline, from vertex to chin; encircle the mouth at the margin of the lips. . . . First, reflect the skin between the eyebrows and vertex. Notice that the skin is closely adherent to the thick and tough subcutaneous fascia. Leave this fascia intact. Nerves and vessels run [through] it . . . the skin of the face is thin. There may not be a considerable amount of subcutaneous fat. Reflect the skin carefully. Do not damage the underlying pale and inconspicuous facial muscles. Observe the thin and loose skin of the eyelids. Remove the skin at the margin of the eyelids. Reflect the skin of the face inferior and parallel to the inferior border of the mandible."

Sherry, Jen and Ivan observed Udele in silence, speaking rarely and then only in whispers, as if not to break a spell. To prevent the undertaking from seeping into the subconscious—and thereby fulfilling Leslie's prophecy—the discourse never strayed from the clinical aspects of the act in which Udele was so ardently engaged.

It soon became evident that the only thing more unnatural than cutting open a human face was for the team at Table 26 to lapse into prolonged periods of taciturnity. Deflecting attention from the particularly disturbing procedure, Udele rose to the occasion by initiating a conversation about, of all things, children.

"I think I'd like three or four," she ventured.

"Spend an afternoon at my house," Sherry responded.

Normally anxious to rack up as much scalpel time as possible, Sherry hung back during the initial dissection of the face. When offered the scalpel, she amenably turned the overture aside, blaming her reticence on the uncharacteristic nausea that had settled over her the previous day while working on the neck. While Sherry attributed the queasiness to stress and exhaustion, her demeanor in relating an experience earlier in the week indicated that perhaps another factor was in part to blame.

With the dissection of the face on the agenda, on Tuesday morning Sherry had taken advantage of a final opportunity to bring Jason to lab to sate his curiosity about the chapter in his mother's life that brought her home each night exhausted and smelling foul. Arriving with Jason first thing in the morning, Sherry found the lab empty of all but the bodies, providing the opportunity for her to ease her son into what he was about to witness. The fourteen-year-old had been transfixed, fascinated and, as far as he let on, unaffected by the experience. That night at home he barely mentioned it.

Curiously, the encounter lingered with Sherry, who, for the first time, glimpsed the cloistered world of a first-year medical student through someone else's eyes. "It really put everything into perspective," she explained to Jen.

"People just don't understand what we do in here or how this works. I had to explain to my brother the other day that we don't put him [the cadaver] in the freezer every night," said Jen, painstakingly attempting to remove the translucent skin from the face without disturbing the subcutaneous structures residing below.

Sherry's confession along with Jen's observation passed without further comment. The emotional component remained an issue untouched for those at Table 26, even in the midst of an exercise that severely tested the psychological endurance of many of their classmates. Abiding by an unspoken edict, in the company of one another, Udele, Sherry, Ivan and Jen continued to keep in check whatever emotion each of them might have been feeling about what was transpiring on the table.

As they systematically stripped the skin from the faces of the cadavers, others weren't quite so successful at suppression. At Table 28, Kevin DeBrancaga violated the edict with such frequency that, after listening to him vent about robbing the cadaver of his identity for an hour, his lab partners finally told him to shut up. The people working with Leah Schreiber were much more tolerant of Leah's tendency to recoil during the facial dissection until, for no particular reason, the elimination of the fascia around the eyes transformed her into the picture of clinical detachment. "I think we'd feel like we were taking away her humanity, too, if only we could see one damn thing in here," she cracked.

The person destined to be most disturbed by the activity in lab that afternoon would remain forever a stranger.

"Excuse me, can you direct me to the cafeteria?" she'd asked innocently, approaching the first person she saw upon walking through the metal doors into Lab B, a medical student with scalpel in hand. As the perplexed student began stammering directions the woman's attention was drawn to the source of his consternation, a human face well on its way to no longer being a face.

"I-I-Is that what I think it is?" the visitor sputtered. Two other students moved to her side, spun her around and ushered her out the door. A moment later they returned, wondering if perhaps it might not have been prudent to direct the unexpected guest to the psychiatry department for counseling. The next afternoon, when the Class of 2002 entered Room B527, they did so through a door boasting a hastily penned sign reading LAB.

The source of the woman's discomfort could at least be explained, which was more than could be said for Sherry's reluctance to assist her lab partners in such a critical phase of the dissection. Of all people, it made no sense for someone who'd spent her entire adult life encountering death at its most appalling to suddenly become squeamish.

Nurses working emergency rooms in New York and Los Angeles quickly learn the value of detachment and the consequences, in terms of career and psychological well-being, for those unable to relegate the daily onslaught of abomination to the recesses of the mind. More often than most medical professionals care to admit, there exists the situation, the person, the catastrophic circumstance that manages to snake its way into the crevice, destroying every coping mechanism in its path.

Sometimes, as was the case during Sherry's assignment to the University Hospital organ transplant team, coping became a simple matter of understanding what worked for her. As a nurse-anesthetist, Sherry's responsibility once surgeons had harvested the organs was to disconnect the patient from life support. Although the task was easy to rationalize intellectually—after all, the patient's final act would perhaps save another life—Sherry still found it easier to absent herself once she pulled the plug, opting not to be there when the cerebral monitor flat-lined, mutating the interminable chirps into a lifeless monotone.

Out of self-preservation, Sherry had long since stopped counting the number of people whose final earthly vision was that of her face hovering over them in the moment before they went under. Still, the knowledge was always there. "When you put someone to sleep and you look into their eyes and then something happens and you realize yours were the last eyes they looked into before they died, it changes you. It has to," she said.

The eyes forever imbedded in her conscience were those of a young man, no more than twenty, shot in the gut, dispatched immediately from the ER to the surgical trauma unit. He had red hair. Along with everything else, Sherry would always remember that; never before had she seen a black man with red hair.

His vital signs, strong when they wheeled him into the STU, were buttressed by a level of circumspection profound for someone bleeding profusely from a gaping wound in the abdomen. Most of the time, Sherry was lucky to get a grunt or a groan; this guy was positively loquacious, insisting on telling the nurse his life story. He'd just finished the part about being paroled from prison two days before when Sherry signaled the end of the monologue by holding aloft the translucent mask.

His face folded into creased consternation. "Am I going to die?" he asked, a question asked her many times. She provided her stock response: "Of course not. No one ever dies on my shift." With the monitors showing his heart rate and blood pressure still stable, Sherry had no reason to believe otherwise. Beneath the mask, she thought she saw the trace of a smile.

Then all hell broke loose. The incision unleashed a torrent of blood, causing the gauges to spiral precipitously downward. As the surgeons worked frantically to staunch the internal bleeding causing the hemorrhage, detachment abandoned Sherry. "You can't let him die," she implored. "You can't. I told him he wasn't going to die."

By then, he already had. Afterward, Sherry made a personal vow to promise nothing, even glibly, whenever a patient invoked the question of mortality.

She thought it strange that a moment so defining occurred just a few days after she'd been accepted into medical school. Placing it in the context of prophecy, she drew comfort from knowing that, confronted with parallel situations in the future, she would be an active interventionist and not a passive observer.

The months between the call from George Heinrich and her first medical school class the following August were interminable. After finally achieving the singular goal on which she'd expended so much forethought, Sherry couldn't wait to get started. Once classes began, she was already projecting ahead seven years. The ER, the operating theaters, those were but way stations in the hierarchy of the health care system. At the end of Sherry Ikalowych's odyssey there waited a bustling medical practice, an office where she would have her own staff, her own patients; she was going to be a confidante, a counselor, and when she gazed into the eyes of those patients it would signal not the end of their consciousness but the start of a dialogue.

She couldn't have cared less that it all sounded hopelessly idealistic. After twenty-three years spent observing the disparate parts of medicine, Sherry Ikalowych was ready to put the puzzle together.

Overcoming her perturbation, without ever really understanding if the source was nausea or Jason's visit to the lab, Sherry accepted the scalpel from Udele as the final slivers of primary fascia were removed from the cadaver's face. With the probe forceps, she began searching for the facial artery and nerve. The limited amount of space, coupled with the congestion of secondary fascia and facial muscles, made delineation extremely difficult.

"Can't tell what's fascia and what's not," she said.

"Maybe if we clean around it," Udele suggested, a recommendation that immediately paid the afternoon's only anatomical dividend before the beginning of the long Presidents' Day weekend.

When they returned five days later, the ubiquitous first-year smell seemed oddly comforting and familiar. Only Udele and a few others still bothered with the ritual of warding off the odor by double-gloving and squeezing soap between the layers. Wreaking of phenol and decaying human flesh became a totem: Asked by an opponent in an intramural basketball game which of her teammates bore the stench of gross anatomy, Jen had replied defiantly, "We all do," and hoped the redolence could be used to ATP/2002's advantage.

Jen had spent the weekend visiting friends in Boston, and while the four days had provided a much needed break, she found it impossible to totally adjourn from the lab. Watching television with a friend, Jen had lain her head upon his chest and "right away, I thought, 'I'm lying on his PMI [point of maximum impulse],' and then I thought, 'No, wait. That's a little bit higher.' And then, when I was talking to my friends, or I saw someone standing on the street, I started envisioning what's inside. It's like now I don't see people anymore, I see their anatomy. Scary."

"That's good," Sherry countered. "It means you're starting to look at the body aesthetically."

Confronted with the infinitesimal nerve and circulatory systems embedded in the face, the deliberate pace established by Sherry began to pay dividends as Table 26 managed to preserve more structures than they destroyed. Even so, Jen wouldn't have minded if Sherry had prolonged the laborious examination along the cheek and jawline just a bit longer.

"You're getting into my danger zone," Jen warned when the dissection around the eye arrived sooner than she would have liked. Sherry ignored her and continued to poke under the eye sockets with the forceps, searching for the orbicularis oculi muscle and the infraorbital nerves. Buying into the "windows of the soul" theory, Jen had long been apprehensive about the eyes. Yet when Sherry handed her the forceps, Jen accepted the instrument without hesitation and picked up right where her lab partner left off, abashed more by her capacity to set aside the typical human response than by the deed itself.

The ability to transcend emotion also served Jen well at night. On the rare occasions when she dreamed of Number 3426, her subconscious never once conjured anything even approximating a nightmare. Thus, in the evenings that followed the procedure she'd most dreaded, Jen was spared the nocturnal vision of the classmate who felt compelled to share it with the rest of the class via e-mail:

"I'm having dinner with my mom and there's this huge eye model on my plate. Yes, the eyeball with all its juicy muscles and nerves. It was even labeled. Then my mom picks up the eyeball and tells me she's glad the superior oblique is on the side (???!!!). I'm staring in disbelief, and she asks me to go ahead and eat. I shout in desperation: 'But why did they have to label it?!' I wake up to the cry of my alarm clock. Thank God! Please don't analyze this dream, I'm really not that sick in the head!"

Jen's ability to shove aside the psychological impasse she'd expected to encounter around the eyes was, as much as anything, a product of the frustration that came from operating within a limited milieu. "The *Atlas* makes it seem so easy to find this stuff, but once you get in there you can't find anything," Jen complained fifteen minutes into a thus far fruitless search for the infraorbital nerve.

Looking ahead to the exam, Sherry examined the morass that once was a face and wondered: "How are they going to tag this stuff? There's not going to be anything left to tag."

Overhearing the exchange, Shamshad H. Gilani, one of the two instructors rotated into Lab C for Unit II, offered advice to Table 26 and, by dint of raising his voice, to the rest of the lab: "*Dissector* first, then dissect. *Dissector.* Then dissection."

"I *looked* at the *Dissector* first," Jen protested under her breath, placing the scalpel in Udele's palm, hoping her lab partner would find success where she had failed. Assuming the quest for the infraorbital nerve, Udele tugged delicately on a minute white thread, separating the surrounding tissue to provide Jen a closer look.

"I think I see it," she murmured.

"Where? Wait. I think I see it, too," said Jen. "Right there. It's coming out of the, oh, what's that part of the eye called? Damn."

Succeeding at a task that had stymied everyone else at the table, Udele continued to snip and prune around the nerve with single-minded concentration.

"I still don't have the feel yet, that's why I let [Udele and Sherry] do most of the cutting. They can feel stuff that I can't feel yet," Jen said, envious of her partners' instinct for sorting through extraneous tissue.

Udele looked up. "Not really, it's tough. Especially here," she said quietly, gesturing toward the face.

As with the arm and hand, the face provided ample opportunity for clinical alliance with the anatomy. That translated into another impromptu lesson from Sherry, who used the path of the mental nerve as it snaked through the jaw to explain how a shot of Novocain from the dentist managed to simultaneously numb the chin and lip.

Elsewhere, the faculty engaged in a few demonstrations of their own, a favorite being the manipulation of the orbicularis oris muscle to demonstrate how human beings, from infancy through senectitude, perpetuate the universal sign of affection. When a few years before an instructor had importuned that "This is the muscle you use to kiss your girlfriend" the student to whom he imparted the information shot back, "My wife, too?"

In Lab C, Paolo Varricchio's ministrations extended beyond the relationship between facial muscles and matters of the heart. Trumpeter Dizzy Gillespie, he explained, owed the signature ballooning of his cheeks to the buccinator muscles. For those not readily familiar with Dizzy Gillespie, and a few clearly were not, Varricchio gave an impromptu demonstration, dramatically puffing out his own cheeks. The analogy to Gillespie was apt, he continued, because buccinator is the name of the long trumpets used to herald royalty in the Roman Empire.

Sensing an unease among students accustomed to receiving perfunctory answers from anatomy instructors, Varricchio apologized for his verbosity. "I try to make it interesting; if it bothers you, please let me know," he told them.

In the unit acknowledged as the toughest segment of gross

anatomy, most of Lab C felt they'd caught a break in Varricchio's assignment to their room. Combining levity with common sense, Varricchio brought to the lab a healthy dose of relativity along with a corny sense of humor that often resulted in unfortunate puns like "People are dying to get in here . . . and once they do, they rest in pieces."

A gynecologist in his native Italy, Varricchio regaled the students with the story of how he'd delivered two babies in a parking lot, the only illumination coming from automobile headlights, after an earthquake forced the evacuation of the Perugian hospital where he was employed. Varricchio loved gynecology, but he had an innate sense he'd love teaching even more, and in 1987, three years after coming to America to be with his New Jersey–born wife, Anthony Boccabella brought him to Newark as an adjunct instructor. Ironically, the department initially assigned the former OB/GYN to teach dental students, a collective that devotes twelve weeks to learning the anatomy above the shoulders and only two weeks to what lies below. The apprenticeship served Varricchio well: "That's why I'm so comfortable with the head and neck; that's all I did when I first got here."

Try as he might, Varricchio couldn't instill the same level of comfort to the students in Lab C on the afternoon he summoned the room to Table 28 to demonstrate the prescribed manner of cutting open the defining bone in the configuration of the face. "Make sure you go away from here with as many fingers and toes as you came in with," he said, hoisting the electric Stryker autopsy to exhibit the proper way to cut through the zygomatic arch, the cheekbone.

The Stryker ground into the arch, churning bone, spitting it into the air. Instinctively, the gathering at Table 28 took a collective step backward. "When you use the saw, be careful of the eyes and don't breathe when the dust comes up. Remember, it works just like a hacksaw," he instructed, pulling the trigger, causing the lab to rattle like a high school woodshop.

"This is the sort of stuff I want to tell my friends about that they don't want to hear," one student said. "It's just a shame, because this is the class I want to talk about the most."

In ensuing days the Stryker made the rounds of Lab C. When the saw came to Table 26 it naturally wound up in Ivan's hands. Compared with his work on the ribs, the spinal cord and the other manual labor he'd been subjected to by his lab partners, sawing through the zygomatic arch with the 120-volt power of the Stryker was effortless.

The transection of the cheekbone complete, Ivan rejected Varricchio's suggestion that a hammer and chisel, and not the saw, be used to bisect the superior ramus (jawbone), an ill-advised decision with disastrous consequences. Applying too much pressure, Ivan snapped the blade in half. Dejectedly, he set down the saw and waited for Varricchio to learn of the damage; it had been that kind of semester.

Without remonstration, Varricchio replaced the blade and advised Ivan that, if he still insisted on using the saw, to exert minimal pressure. A minute later, as the new blade broke through the bone, Udele's reward for leaning forward to assess the situation was a huge bone chip lodged in her hair.

"Better than fat, just ask Jen," Ivan said wryly.

The ashes dotting the foreheads of several classmates reminded Sherry that she needed to stop at church on her way home. Before she could receive the Ash Wednesday sacraments, though, the matter of ever more muscles, nerves and blood vessels presented itself.

"I was actually more surprised at how many muscles there were in the forearm," Jen said. "Think about all the contortions we can do with our faces, but the arm—remember all those muscles? Maybe the guys weren't surprised because they're more aware of arm muscles, but I know it surprised me."

Departing from the tradition of instructors who hung in the corners of the room waiting to be summoned, Varricchio was a ubiquitous presence, dashing from table to table, displaying a telepathic capacity to answer questions before they were asked or, too often, a penchant for raising issues of no consequence to the students.

Interested not in the least by what occurs when a crushed ethmoidal sinus under the eye severs the olfactory nerve, Table 26 received a clinical lesson on the matter anyway. "Lots of boxers can't smell because the ethmoidal is smashed," Varricchio explained. "And that's why we wear seat belts, even though they cause whiplash. The olfactory nerve goes and you can't smell anything." He paused. "Then, of course, that qualifies you to work in an anatomy lab."

Failing to evoke even a slight chuckle, Varricchio began to elaborate on the function of the cribriform plate until, noting that Ivan and Jen weren't attempting to pay attention, he moved over to Table 27.

"Cribriform? Where's that?" Sherry asked.

"Don't know; every day I can only absorb so much," Jen responded.

"It's like a bucket: You can keep putting water in, but eventually it's going to start spilling out," said Ivan.

No water was added to the bucket that evening as nearly the entire class, Sherry excluded, remained at school for the annual NJMS *Follies*. Promoted by the student government association as "a night of beer, pizza and intrigue," it better fit the description of a school administrator who called it "five solid hours of sick, med school humor." The event's biggest stars were the five first-years who dropped their boxers to reveal, à la *The Full Monty*, the year's surprise cinema hit, the letters *UMDNJ* scrawled across their gluteus maximus, as *Grant's* described it.

That the *Follies* didn't conclude until nearly midnight was of no particular concern to the Class of 2002. Under normal circumstances, attendance in the lecture hall on Friday mornings was minimal as the class increasingly relied on transcriptions of the copious notes taken during each lecture by the designated scribe.

Beyond the hangovers the students expected to be nursing in the morning, the anatomy department had provided reason for them to stay home all day since, for the first time all semester, the lab would be closed to allow the staff to prepare the forty-five men and women resting in B527 for the next stage of the dissection.

CHAPTER 13

———————•———————

Most of the class had yet to roll out of bed the next morning when Paolo Varricchio, David Abkin and Roger Faison entered the lab, their hands sheathed in heavy-duty rubber gloves, their bodies covered by full-length surgical gowns. Protected also from forehead to chin by transparent plastic face shields, the trio could have been mistaken for a Haz-Mat squad dispatched to clean up a contaminated nuclear facility. In his right hand Faison carried a Black & Decker wood sander retrofitted with a saw blade.

Prior to departing lab the previous day, each table had been reminded to prepare the scalp for a craniotomy. Truth be told, there wasn't much to get ready: any hairlines that hadn't receded because of advanced age had pretty much been laid to waste during the dissection of the face.

With good reason, the department of anatomy warned the students away from the lab on the day the craniotomies were performed: The act of sawing through the skull in an orbit that began and ended just above the eye sockets was an act that repulsed Varricchio, Faison and Abkin, three men with a capacity for disgust totally off the charts.

Once, the students themselves performed the task with hacksaws, a practice that ended when the supply of new blades couldn't match the demand caused by broken ones. With two, sometimes three days devoted to craniotomies, the anatomy department concluded it was too expensive in terms of time and money to allow the students to continue. Once the department turned the craniotomies over to Faison, he jerry-rigged the hand sander in order to expedite an assignment he found despicable.

The mortician's aversion was reflected in his facial contortions as the trio moved through the lab, observing an identical protocol at each table: Varricchio and Abkin holding the heads in place as Faison

197

applied the Black & Decker to the target. The metal slicing through the skulls puddled dollops of embalming solution and spinal fluid, thickened by brain matter, onto the stainless-steel tables as the stench of singed flesh and bone, coupled with the metallic odor of an over-heated electric motor, turned fulsome the level of acridity in rooms already fouled by decomposing bodies.

"Disgusting," said Varricchio, his voice muffled by the shield. "But what are you going to do? These skulls are a lot like the students: Some have hard heads, the others are hardheaded." Averaging two minutes per skull, a pace Faison credited to a new graphite blade, the trio worked straight through lunch, finishing in the early afternoon. Based on past experience, Faison predicted that three days hence the students would return to lab appreciating neither the service they'd performed nor the aesthetics of their handiwork.

Past experience had not included Jennifer Hannum, who fairly bounded into lab Monday afternoon proclaiming, "This is the moment I've been waiting for. I've dissected hamster brains, rat brains, seagull brains. Now, finally, the real thing. I can't wait."

Determining the craniotomy, an incision circling the top of the skull, to be the starting point, Udele, Sherry, Ivan and Jen pondered the next step, their discourse rudely interrupted by a resounding thud halfway across the room. "Whoa, somebody already got their skull off," said Jen, until someone reported the percussion to be a block of wood, used to prop up the cadavers, that had fallen to the floor at Table 28.

In contrast to the tables that at the end of each afternoon fastidi-ously positioned the towels covering the cadavers while aligning the scalpels, probes, scissors and forceps as though they were arranging a place setting for a state dinner, Table 26 wasn't even in the running for the lab's good housekeeping award. Their habit of pretty much leav-ing things as they lay, which had nearly cost them the *Grant's Dissec-tor* before the first practical exam, had, by the midpoint of the semester, resulted in an unkempt table cluttered with wadded paper towels, used scalpel blades and an accumulation of tools that they'd neglected to return to the supply room.

"You have *two* chisels? How did you get two chisels?" Varricchio asked, recognizing the consequences should David Abkin ever learn of the overdue tools.

Ivan feigned surprise. "We have *two*? I don't have any idea how they got here."

"Don't have any idea. You sound like my son." Varricchio smiled, motioning for Udele to insert one of the chisels into the skull fissure created by the Black & Decker. As he began to pry at the front of the skull with scissors, Varricchio directed Ivan to insert the other chisel—as long as it was there, it was just as well put to good use—into the incision at the rear of the skull.

"OK, now what you want to do here is take off the cereal bowl," Varricchio grunted, describing perfectly the top of a severed human skull turned upside down. The bone separated with a crunch and the cadaver's head rocked backward precipitously.

"Eeee-yewww, that sounds so nasty," said Udele.

"Oh no, his whole head is coming off," said Sherry.

"Man, he must have a really bad headache," said Ivan.

"This, truly, is the best," said Jen.

Udele started trimming the dura mater, the cushion separating the brain from the skull, paying no attention to Sherry's account of a previous day so entirely devoted to studying that she never bothered to change out of her pajamas. Intensely tracking Udele's every move— "When you're good, you're good; I'm not messing around with anything she's doing"—Jen ignored her other lab partner as well.

Ten minutes later, Udele effortlessly separated the cereal bowl from the lower portion of the skull.

"Wow, look how nice you did that," Jen enthused. "This is so cool."

His mouth agape, Ivan resembled a Little Leaguer visiting the New York Yankees clubhouse.

"Just like the book, we have a perfect head," Sherry proclaimed, immediately undertaking the search for the meningeal artery.

Without decree, the table became lost in thought, unmindful of the cracks and crunches as other tables pulled apart their cereal bowls.

"I love the brain," Jen whispered, breaking the reverie. "It's the one thing we don't know everything about. And we're using our own brains to try to understand this brain. It's kind of weird."

"An enigma, a mystery," Ivan added.

"A conundrum, the final frontier," Sherry agreed. She looked around the room. "How did the others get their brains out so fast?"

"They ripped them out, but we wanted to preserve the integrity of ours," Jen explained.

For the first time all semester, Ivan took the initiative during a milestone procedure and began to extract the brain, scrupulously slicing through the dural folds toward the brain stem, heeding instructions to maintain the integrity of the arterial system known as the circle of Willis.

As Ivan unhinged the cerebrum centimeter by centimeter from the dural folds, Jen stepped up to cradle the organ. Totally unaffected, looking for all the world like someone who held a human brain in her hands every day, she disregarded the fluids dripping through her gloved fingers to blithely detail the attributes of her neighborhood delicatessen, the whereabouts of friends from Rutgers and whatever else popped into her mind.

"So, yesterday I was studying," she chattered as Ivan labored on, "and my ring finger and my small finger went to sleep. And I realized I was leaning on my ulnar nerve and I couldn't stop thinking about what it looked like, you know? Remember?"

"That was the last unit. I don't have to think about that anymore," Sherry said gratefully.

"Excuse me," said Ivan, "are we supposed to cut the brain stem at the base or lower?"

"Lower, I think," said Jen.

His arms heavy from the intricate finger work necessary to preserve critical tissue, Ivan offered Jen the scalpel; Jen deferred to Udele, who, within a minute, stopped abruptly.

"What's wrong?" Jen asked.

"Nothing, I just think I'm cutting through stuff I think we might need later," said Udele.

"Don't worry, we're not dissecting the brain once it comes out," Sherry reminded her (the complete study of the brain would come later, in a neuroscience course), "and it's not like you're going to make him forget his fiftieth wedding anniversary."

"Whoa, what's going on here?"

Udele froze as Vasan nudged into the table's prime dissecting position. "Cut out the brain stem, anywhere, that's all you need to do. It will come out," he ordered. Ivan gave Jen an I-told-you-so look as the course coordinator moved to Table 28, where hands were waving desperately in the air.

With Ivan positioning the head and Jen continuing to nestle the brain, Udele severed the brain stem in a single motion. "God, this is like giving birth," Jen whispered as, gently, she lifted the cerebrum from the skull.

Some brains came out intact, others all but disintegrated ("I keep wondering what memories I'm holding," said one student as the tissue dripped through his fingers); Table 26's brain broke into five distinct pieces.

"Shit," said Sherry, plopping the parts into the cereal bowl. Brain matter covered everyone, especially Jen and Udele, who exchanged saturated gloves for new ones, tacitly acknowledging that the brain, though fascinating and enigmatic, had severely tested their tolerance for the macabre. "It is just too gross," said Sherry.

Jen looked at the brain lying disparately in the skull. "I'm so disappointed. I'm afraid it's going to turn to mush and I really wanted to hold it." She wandered off and returned a minute later with another brain. She placed it on Number 3426's chest.

"Look, from those guys." She gestured to Table 30.

"A perfect brain," Sherry said in awe.

"You can see the circle of Willis," Udele added, equally enthralled.

Using *Netter's Atlas* as a guide, Sherry pointed out the pertinent structures. "I am so jealous, this is *so* beautiful," Jen sighed.

"Switch it tonight," recommended Vasan, shouting over his shoulder from Table 25.

Visibly upset, Jen caught the attention of Varricchio, who commiserated by explaining that often the embalming fluid doesn't penetrate past the spinal column. "Besides," he added cheerfully, "the brain is always the first organ to decay."

"But I just wanted to hold it," Jen repeated, the picture of dejection.

With less than two feet separating one lab team from another, the exhortations of one table, good or bad, became the business of all other tables. Thus, by the end of the afternoon, Table 26's lamentations were well known to everyone within earshot. As Jen began the ritual of cleaning up for the day, Persephone Jones, from Table 24, interrupted her.

"It's OK," said Persephone, an empathetic big sister reaching out to a sibling with a broken toy. "You can play with our brain if you want."

Jen smiled meekly. "Thanks," she replied gratefully.

The removal of the brain left the cadavers, bizarrely, with empty heads, a phenomenon that allowed anyone so interested to quite literally look at the world, or the lab anyway, through someone else's eyes. Peering into the void, the students could see sinuses that tormented the allergic; the canal leading to and from the ear; the palate; and, underneath that, the tongue. For an entire day, the Class of 2002 performed the inside-out inspection, memorizing the foramen, the entry and exit points for scores of nerves and blood vessels. Tiring of that exercise, they then compared the bone structures found inside the skull with radiographs posted on the light tables.

Table 26 proceeded systematically, going strictly by the book or, as was the case, syllabus, an investment of intellectual capital that, not coincidentally, obscured an issue of greater magnitude.

In the ongoing conversation that ran the whole of the semester, Udele, Sherry, Jen and Ivan flitted between the mundane—shampoo preferences, roommates, personal relationships—to the clinical terminology that sets physicians apart from the patients who would one day see their care, comfort and expertise. Since January, exegesis had emanated from an array of sources: the cadaver, textbooks, the faculty and from one another. Though confronted each day with a miracle, not once had there been even tangential mention of the larger scheme. The closest they'd come had been during the brief incursion into evolution while discussing the fate of the palmaris longus, the muscle generally missing from inside the arms of modern human beings. At that, no one thought to mention, let alone acknowledge, the role of a higher authority.

On the surface, Table 26 certainly didn't seem averse to such discussions. Despite the constraints imposed by family and school, each weekend Sherry usually found time to attend Mass. While not quite so observant, Ivan had also emerged from a strong Catholic tradition. Udele prayed each day for guidance and strength. Lacking a formal religious upbringing, Jen nonetheless had an innate curiosity about spirituality.

And yet all kept private their thoughts on whether the mortal remains they deconstructed each afternoon were divine in origin: Udele was too private to openly discuss her faith; Jen much too unsure about her own position; Ivan's thoughts were on his grades, not intellectual forays more suitable for a philosophy class than a medical school; and Sherry, after years spent observing the dichotomy of failure and resiliency in the human body, was far too pragmatic to bother with a question for which she was certain there were no answers. "With everything else going on, who has time for that?" she wanted to know.

Bradd Millian, for one. Although Bradd expended not an inordinate amount of thought considering the possibility that the fifth member of the team at Table 39 was perhaps present in more than body, the idea did occasionally cross his mind. Medical school being a hotbed of twenty-two-year-old Darwinians eager to eviscerate any suggestion that anatomy is anything but the osmosis of a single cell emerged from the primordial soup, for the better part of the semester Bradd maintained a steadfast silence on the subject.

As his table audited the tissue entombed within their cadaver, keeping his mouth shut wasn't always easy. Given Bradd's nature, inevitably an inquisition such as theirs yielded to a larger question. Finally, one afternoon, Bradd asked it:

"Do you ever wonder if he's watching us?" he asked his lab partners. By the looks on their faces, he could tell they weren't exactly certain who "he" was. Bradd nodded at the cadaver. "Him. You know, do you ever imagine he's looking down on us, watching what we're doing and thinking, 'Wow, man, don't do that?'"

Having gained everyone's attention, Bradd continued, "I think that he thinks it's OK. I mean, it's what he wanted. He donated his body for this. I'll bet he's looking down, thinking, 'That's pretty cool, what they're doing.'"

Fortunately, in his lab partners, Bradd had a receptive audience. Given the number of students embracing the position of Leah Schreiber—"What they see as an engineering marvel, I pass off as billions of years of evolution. This is a product of evolution, plain and simple. I don't think God put this together"—he might just as easily have encountered a hostile reaction.

In this de facto separation of church and state, the evolutionist majority rule made it clear, quietly yet firmly, that issues interposing the existence of a higher power were best not communicated. In the med school version of don't ask, don't tell, Charles Darwin was rarely mentioned and, while God's name certainly got more airtime, the context almost always constituted a violation of the Fourth Commandment.

Because the subject was never broached, few in the Class of 2002 had any awareness that, within their subculture, others had clear and well-formulated ideas about the relationship between God and the divinity of the human body. They didn't know that Nagaswami Vasan listened to taped prayers in his car on the way to school, prayed to the mantra tucked behind the door of his office at noon each day and had yet to participate in the semester when the basic tenets of his faith weren't challenged and then ultimately reinforced by what he witnessed in lab. Nor did anyone have a clue that Paolo Varricchio, the wisecracking instructor who projected an air of blasé indifference to the cadavers, was a devout Roman Catholic given to unabashedly pronouncing, "The more I understand how complicated and balanced the body is, the more I see how marvelously the works of God are put into practice."

Varricchio well understood the dichotomy of that observation: A student of anatomy, historically as well as clinically, he understood that the honor of God, and not the deficiency of human intellect, for centuries represented the single greatest impediment to the advancement of modern medicine.

Ultimately, the two men credited with vaulting anatomical investigation into the modern age were not, strictly speaking, men of science. Between them, it is estimated that Leonardo da Vinci (1452–1519) and Michelangelo (1475–1564) performed as many as five hundred human dissections, all sanctioned by a Church persuaded that the accurate portrayal of saints and, in fact, Christ himself could only be accomplished through the visual documentation of human anatomy. The lifting of the Church's sanctions allowed da Vinci and Michelangelo to present to the medical community schematics of the skeleton, the muscles, blood vessels, sinuses, heart and brain that, for all intents and purposes, are as relevant today as they were four hundred and fifty years ago.

"Had [da Vinci] written an anatomical text, as planned . . . the process of anatomy and physiology would have been advanced by centuries," wrote Charles Joseph Singer in *A Short History of Anatomy and Physiology from the Greeks to Harvey*.

Although this is a bit of an overstatement since da Vinci was, in fact, guided by Galenic principles, as it happened just five years after da Vinci's death, Andreas Vesalius forever changed the course of medicine.

Just ten months shy of the third millennium, the moment when Bradd Millian sensed an extrasensory spirit at Table 39, the pendulum that had begun its arc over two thousand years before in the sphere of philosophy and religion had swung full into the domain of secular evolution. Not everyone in the Class of 2002 held *Origin of Species* in higher regard than the Book of Genesis. But enough did that it was somewhat aberrant for three young people assigned to one table to similarly ascribe the origin of life to a higher entity.

Inspired by Bradd, Gayatri Rao and Elizabeth Garcia began discussing an issue either ignored or studiously avoided by most of their classmates. Despite bringing three hugely different perspectives to the table, Bradd, a Jew; Gayatri, a Hindu; and Elizabeth, a Christian, had each independently reached a conclusion identical to that of Varricchio: Seeing the anatomy, touching it, cutting it and, in some instances, destroying it was a transcendent experience that sustained and strengthened but never diminished their faith.

"It really does make you wonder what happens after [death]," said Gayatri, at twenty-one the youngest member of the class. "You look at what we're doing here, at what happens to the body, how it works, and spiritually you have to believe that the person has gone somewhere else. That's comforting, it's nice to know."

Elizabeth Garcia concurred: "It's like this is his house, but he's not here anymore. And we're just sort of remodeling."

Unlike the evolutionists, who tended to be absolutists, Elizabeth,

Bradd and Gayatri parlayed flexibility into the dogma, acknowledging evidence that the anatomy had, indeed, transformed itself over time.

"If you take it as far back as the apes, OK, you have to say we've evolved. But we had to have started at some point for which there is not a scientific explanation," said Gayatri.

"It's part of a master plan, but the master plan is not the end product," Elizabeth added.

"Right, this really might have started out as primordial soup, but there had to have been something before that, maybe not something that everyone agrees is plausible, to make it possible," said Bradd.

"I really do think the body is proof of a higher authority; there has to be a God or, depending on what you believe, someone in heaven who is up there overseeing this," said Gayatri. "It scares me enough to believe that there has to be something more than this."

"I do believe it's something or someone, but I can't say for certainty what it is," Bradd acknowledged.

"You take everything we've learned, starting with single cell biology, the development of the amino acids, all of it. What are the chances that all those things would come together to form a functioning, cogent human being? Why can't it be both? Why can't it be part creation, part evolution?" Elizabeth wanted to know.

Had he been part of the conversation, Paolo Varricchio would have weighed in that, perhaps, there is common ground between secular humanism and creationism. Discussing his own beliefs one morning prior to the start of lab, Varricchio allowed, "We're still evolving from the dinosaurs. Every day we see it, muscles disappear, organs that are useless. We have no idea what nature has going for us down the road. My father, who wanted to be a priest, once told me not to use science to limit my religious beliefs. And I don't. God created life, but God created evolution, too. Who is to say that the Big Bang is not an act of creation?

"Religion starts where science ends. Without creation, there would be no life. I really believe that. Life would be at odds with itself. Without belief, you can't bridge that gap between what happened and what science can't explain."

To explain what couldn't be articulated without taking a leap of faith required closing several other chasms, most notably the one thrust most often at even moderate creationists by their opposite number: If the human model is God's perfect creation, what, then, of the lower forms?

Bradd himself raised the question and then answered it. "You really do have to ask yourself if the rat, too, is a perfect specimen. After all, if you believe God created man then you have to believe God also created rats. You can't have it both ways. And then you have to ask yourself, what makes the perfect human better than the perfect rat? Do rats go to heaven, too?"

Elizabeth and Gayatri looked at Bradd and then the cadaver. Although the philosophic discourse had up to that point been invigorating, Table 39 had a priority beyond deliberating the afterlife of rodents. Chief among those was completing their analysis of the cadaver's skull because, come the next day, they would pretty much destroy the very component that had moved Bradd to incite the conversation in the first place.

The procedures as described in the syllabus introduction on two consecutive days in late February rang ominous: "Disarticulation, Prevertebral Region & Base of Skull and Bisection of Head and Nasal Cavities."

Translation: First sever the skull from the spinal column and then cut it vertically in half. Faced with an act that, by any definition, qualified as a dastardly deed, the class managed to find a measure of solace in the knowledge that, unlike their predecessors, they wouldn't actually separate the head from the body. Still, even the most cursory review of the *Dissector* made it pretty obvious that, by the time they finished, what remained of each cadaver's head would be left hanging by only a thread.

The disarticulation proved to be disturbing enough, causing nearly every table to draw an involuntary, heart-stopping breath as the chisel-induced separation pitched the skull forward onto the chest.

Little did they know, although some suspected, that barely twenty-four hours later the bisection would make the disarticulation seem bland by comparison.

"I don't think there is anyone in here who isn't freaked. This is weird. I definitely have to say that this is weird," Jen admitted, watching as Ivan, using the hacksaw, cleaved in half what remained of the face and skull.

"I feel like we're taking away his personality," said Ivan, taking a breather.

Beginning with the offer on the first day of lab to purchase extra copies of the *Dissector* with his credit card, Reuven Bromberg had served in the capacity of self-appointed ambassador between the class and the faculty. Once or twice a week, Reuven made the rounds of the four labs, checking everyone's progress, seeing if any conflicts needed to be brought to Vasan's attention. For the most part, no one could say for sure if Reuven's sojourns were the product of boredom, genuine interest in the well-being of his classmates, disenchantment with his own lab partners or infatuation—Reuven, it was generally acknowledged, had a crush on a woman at a table in Lab C.

Whatever the reason for his periodic appearances, Reuven's visits were nearly always convivial. Which explains why the room was more than taken aback when, appalled by the bisection, he turned Lab C into a soapbox, declaring: "This is where I draw the line. I can't take it."

Making an exit worthy of an actor playing the role of spurned lover, Reuven pivoted and walked out the nearest door. Five minutes later he was back, observing the students at Table 30, all of whom were doing their level best to ignore his impassioned soliloquy on gross anatomy as degradation: "We just got rid of whatever was left of his personality; it's all gone now. I think they should be given a choice, they should have the chance to see what is done to them in here before they donate their bodies. I always said I'd donate my body, but today, for the first time, I'm reconsidering the idea. I'll probably go ahead with it but this, this is too much," he announced, exiting again, this time for good.

To be certain, Reuven wasn't preaching to an unsympathetic audience. Annette Pham expressed relief that her cadaver, a man, had worn dentures. "If he'd had teeth, it would have just added another dimension of humanity to it and this is already bad enough," she said.

Leslie Pooser, as usual, checked in, spreading the word to Table 26 that the middle nasal meatus—Leslie was fairly sure she'd identified the correct sinus—was filled with mucus. "Snot," she elaborated, for the enlightenment of those who might have missed the point.

The bisected nose held a peculiar fascination for everyone, Jen included. Probing inside the exposed nostril with an index finger, she passed along a clinical observation beyond that sought by Vasan in the syllabus: "If you stick your finger in your nose, this is where it would go." Taking the full measure of the human proboscis into account, it didn't go far.

Because a skull cut vertically in half eliminated critical anatomical landmarks, Udele and Sherry were forced to continually push the two sides together to orient themselves. The technique allowed them to expeditiously identify the nasal cavities and muscles controlling the tongue. Locating and probing the foramen, the minuscule openings providing nerves and blood vessels access to the vestibule of the mouth, however, caused problems until Udele hit on the solution of using a wire garbage bag tie, stripped of half its plastic sheath, to probe the fissures.

Following Vasan's suggestion that access and understanding had in the past been better accomplished by students borrowing hollowed-out skulls issued by the supply room, Table 26 dispatched Ivan to visit David Abkin's domain. In the manner of Hamlet pondering the mortal remains of Yorick, they studied every nuance of the supply room skull, ultimately rejecting it as too pristine; for the purpose of identifying the nerves and ganglia—connective tissue between the nerves—the cadaver seemed a superior source.

But not, it turned out, far superior. When Udele's frustration in locating the ganglia in the maze inside the mouth mounted, Varricchio came rushing to her aid before bewilderment spun out of control. The inability to pinpoint the ganglia, Varricchio reassured Udele, had nothing to do with incompetence. Gently scraping the hard palate, the roof of the mouth, he noted that "the ganglion may appear to look like fat. That's because you need to remember that the nervous system is ninety percent fat. That's not to say that fatter people are any smarter than all the rest of us." Not intentionally, he looked at Ivan. Aware of the excess weight he'd gained since the knee injury, Ivan patted his stomach, feigning disappointment.

Varricchio directed their attention to the cadaver's throat. To all the other surgical procedures endured by their first patient, the course instructor announced assuredly, it appeared they could add a tonsillectomy. Launching into another anatomical stream of consciousness, Varricchio segued into a mini-lecture on the adenoid, a structure that, he pointed out, does not exist in and of itself but is instead the enlargement of the pharyngeal tonsil.

"Now," he added, "let's talk about the epiglottic glands. What forms in there?"

Sherry shrugged, followed by Udele, Jen and Ivan.

"Tumors," Varricchio said solemnly.

"Tumors?" Jen asked absently.

Varricchio nodded, staring into Sherry and Udele's stricken faces, unaware that their and Jen's discomfort was not in the least related to the revelation that tumors might form in the throat. Basically, all they wanted was for Varricchio to go away so they could close the table and get on with the business of studying for the next physiology exam.

"S-H-I-T happens." Varricchio shrugged. Unsure what else to say, he headed for Table 33, the only other team remaining in lab.

The moment the instructor stepped away Jen and Udele covered the cadaver, sprayed the towels and made haste to get out the door.

When next they returned to lab, two hours after the completion of the physiology test, no one was inclined to grapple with saliva ducts, tonsils or anything else inside the cadaver's head. Prior to the exam, Ivan had managed only three hours of sleep while Sherry and Jen reported tossing and turning all night.

Serendipitously, twice that afternoon fire alarms emptied the medical school. While the frequency of the false alarms didn't approach that of the first semester, when every day it seemed they were evacuated into the courtyard, it still happened often enough to constitute an annoyance, especially with the temperatures near freezing. Making the best of the situation, during the first evacuation several tables headed into the courtyard with supply room skulls, seizing the downtime to identify the foramen while they waited for the all clear. For many, it was the afternoon's only edification. At the sound of the next alarm, most everyone closed up their tables, grabbed their coats and books from lockers and, instead of lingering in the courtyard, called it a day.

Out of necessity—basketball had become more an outlet than a form of exercise—Jen returned a couple of hours later for the basketball game during which she set the NJMS intramural women's single-season scoring record. To her embarrassment, the officials stopped the game to recognize the achievement.

For the second time that semester, Jen and her classmates passed an ominous towel-covered lump at the front of the room as they trickled into lab the following afternoon. Smaller than the shrouded protuberance they'd encountered the previous month, everyone nonetheless knew what to expect when the shroud was lifted.

"Udele has gloves on," Jen said, conveniently delegating the unveiling to her lab partner. Magnetized, the rest of Lab C converged to get a look at the unimaginable display that had emerged this time from the supply room freezer.

"I knew it," Jen muttered, gaping at a severed head sectioned into three parts, like a pizza cut in half with the other half divided into equal quarters. Fascinated by the cadaver's gold teeth—"I can't believe they left those in there"—Udele paid no mind to the spectacle. Nor did anybody else; even Leslie Pooser passed on an allusion to the potential for nightmares. Matter-of-factly, Udele led the rest of Lab C in scrutinizing the head, searching for structures that might have been destroyed in their own dissections. Gradually, Udele's classmates peeled off to attend to business at their own tables, where the time had come to prepare for the second unit's fast-approaching written and practical exams.

The brevity of the second unit shouldn't have come as a surprise; Vasan, after all, had given them fair warning during his impromptu postexam lecture barely two weeks before. Still, the subsequent tumult—deconstructing the face, extracting the brain, disarticulation, bisection—had created a sensation of propelling the students through a vortex where time simultaneously accelerated and remained inert. Were gross anatomy a roller coaster, they were a little shy of being halfway through the ride. The scariest part was over and they could look back with a measure of satisfaction knowing they'd survived the precipitous initial dips and plunges. And though they weren't yet out of danger, at least they could see—or was it an illusion?—the lay of the tracks ahead.

The two factors blessedly exempted from the second exam, fear of the unknown and a physiology test three days prior to the fact, brought small comfort to Sherry, ever aware of what was being sacrificed on behalf of her dream. To assuage the guilt of being wedded constantly to textbooks, she still set aside each Saturday for Jerome and the kids, a contingency that boomeranged into self-reproach for abandoning her studies. "It will be a miracle if I make it through this semester. I just don't have the time everybody else has to dedicate to this," she admitted, joining up with Udele, Jen and Ivan for the structured practical review.

Depleted by a lifestyle that now averaged four hours of sleep each night and five or more cups of coffee each day, that afternoon Sherry ran into her opposite in Ivan, buoyed by learning that morning that he'd at last scored above the mean on the latest physiology exam.

Slogging from table to table, conferring collectively on the sample questions, the ease with which Ivan, Sherry, Udele and Jen worked together was evident in how effortlessly they reached a consensus and in their enthusiasm to embrace the member of the team supplying a correct answer. The chances were slim that they would ever be best friends or, in the future, even hang out much together; after lab, they all went their separate ways. But during those four hours every afternoon, an unspoken and beneficial bond had formed that sometimes seemed to join them intellectually, Ivan included, as one.

The advent of the second gross anatomy exam translated into another lost weekend for most everyone, Sherry being the exception. But for a brief lab foray on Sunday, she spent the rest of the two days fulfilling her familial obligations. Meanwhile, Jen went back to the Barnes & Noble, where she mingled happily with the non–medical student clientele before returning to her apartment to spend "way too much time" watching the Big Ten, ACC and Big East conference basketball tournaments.

After surviving the first anatomy exam with his head barely above water, Ivan didn't have the luxury of indulging his love for basketball over the weekend, and instead spent it shuttling between the study rooms and the lab, occasionally encountering Udele and Joyce Prophete during his journeys back and forth.

Collectively, Udele and Joyce had made the decision to remain at school for as long as it took to master the material they expected to be presented on the exam. Their commitment kept them on campus until five o'clock Saturday morning, a feat made possible by intermittently adjourning to the nearest corridor for impromptu cartwheels and nonchoreographed dance routines. Returning to school five hours after they'd departed, by mid-afternoon they were scrunched into fetal positions on a carpeted hallway floor, surrounded by the detritus of Burger King meals and half-eaten slices of pumpkin pie, giggling uncontrollably, oblivious to the irony of the role sleep and nutritional deprivation was playing in the process of their one day helping others to live healthier lives.

When early Monday morning Vasan and his apparatchiks gathered to tag the bodies, the routine passed without a single disputed question. At Table 26, Station 27 in the practical, Paolo Varricchio affixed a tag inside the neck: "(A) Identify this nerve and (B) In which ganglion are its cell bodies located?"

After the faculty double-checked the fifty two-part questions and answers affixed to bodies, structural models and radiographs, David Abkin circled the four labs, spraying the exposed faces of the cadavers with water before covering them with paper towels to slow the ever-increasing deterioration of the tissue.

Ivan didn't get to school until shortly before noon: Group A, the covey to which he and Jen were assigned for each exam, were scheduled to take the written test at 12:30 and the practical at 3 that afternoon. Although there were no indications he was about to be visited by another migraine, Ivan's eyes were again reduced to red slits. Sleep the night before had been fitful, filled with half-awake dreams about the muscles of the face. In the dreams, Ivan could not identify a single one. Tapped out on anatomy, as he waited for the written exam to begin, Ivan babbled about his girlfriend's search for a job, Miami and myriad other topics, none having a single thing to do with the contents of the imminent test.

Drifting over to the locker area outside lab to shed his jacket and books, Ivan encountered another of his lab partners with the group waiting for the noon practical to begin. "Hello, Sherry," he said jovially, receiving in return a flutter of grim-faced recognition.

Eyes straight ahead, mouth pursed, Sherry was in a world of her own making, totally disregarding the nervous clamor of the classmates surrounding her:

"Welcome to the end of the world."

"What's with the book? Put it away; if you don't know it already then it's too late?"

"If you believe in God, now is the time to put it in His hands."

Eight excruciating minutes behind schedule, the doors swung open and Sherry entered the practical, confident of receiving a satisfactory grade but nonetheless frustrated by the realization that the demands of her other life meant setting aside the original goal of near perfection. It would take a while, Sherry confessed, before she could accept the fact that she would come out of the most crucial medical school semester with anything other than high honors.

Three hours later, after finishing the written exam, Jen entered the lab and, noticing the unoccupied stool at Table 26, Station 27, figured it boded well to begin the practical exam at her own table. Depositing herself there, she took measure of the questions articulated on the tag affixed inside Number 3426's neck.

Meanwhile, across the room, Ramon Nunez passed the time at a rest station near the switch for the Klaxon horn by demonstrating for a classmate a clinical procedure that would not be part of the exam. "If you're not ready, just press and hold your breath," he recommended, the first two fingers of his right hand pressing into the carotid artery. "It will make you pass out and you'll be eligible for a medical excuse to do this later."

"Funny, just push the button," Roger Faison said, ordering Nunez to begin the proceedings.

The instant the Klaxon horn rang out, Jen scribbled the answers, "inferior alveolar" for the nerve and "internal laryngeal" as the ganglion where the cells were located. Setting down the pencil and closing her eyes, she recalled how many times Udele and Sherry had returned to that structure, drumming its name and function into her conscious until it was impossible to forget.

Gazing around the room, she realized the same thing would probably be true of every other blood vessel, muscle and nerve she was about to encounter. The horn sounded and Jen shuffled to Station 28 filled with confidence and gratitude, knowing that Table 26 was doing it right, precisely as it should be done.

CHAPTER 14

———————————•———————————

The end was in sight, the measure of their progress evident in the mutilation of the cadavers from thoracic cavity to the head; the skin from the abdomen to the toes, still undisturbed by scalpels, reflected how much remained. In five weeks it would all be over.

Though the worst was behind them, some unpleasantness still lay ahead. More than a few were apprehensive about what might be uncovered in the lower gastrointestinal tracts. Then there was the scheduled dissection of the anatomical region described not so delicately by the syllabus and *Dissector* as the anal triangle. Finally, they were bracing for a spate of exams, one every week in either physiology or psychology—the physiology department had even scheduled one for the Monday after the seven-day spring vacation—until the end of the semester.

The short-term goal was to get to the break, three weeks away, after which were scheduled four final days of gross anatomy. Buoyed by the knowledge that the duration of GA would soon be calibrated in hours, the class came to lab the day after the second gross anatomy exam more relaxed than they'd been in weeks.

Not so Jen, who began the semester's final phase by obsessing over the physiology department's vacation buster. "I'm trying to decide if I should do something on break, go somewhere, you know, have a life, or should I hang out and study for the phys exam. If I have a life, I know I'll just feel guilty, but if I study I'm going to feel even more like I have nothing going outside of this place," she said.

"I don't think you can have a life, there's too much guilt involved," said Udele, who could count on one hand the number of Saturdays or Sundays since January she hadn't spent holed up in the library or in a study room.

"I know, but I'm trying not to take myself so seriously. I know we're all overachievers, so I have to keep checking myself," Jen told her, observing Sherry faintly trace an outline of the abdominal incision with a scalpel; the purple marker that had so ably diagrammed the path of every other initial cut lay useless on the table, ruined by the absorption of various postmortem excretions.

Sherry didn't give spring break a second thought—assuming the kids managed to stay quiet before school, her greatest extravagance figured to be sleeping in on a couple of mornings. Her focus was more short-term: her husband's birthday celebration that evening. Jerome's solitary wish was that she get home early, with a cake, so the whole family could sing "Happy Birthday" together after dinner. Given his efforts on her behalf, Sherry thought it didn't sound like much of a celebration, so, after the exam, she made straight for the mall to pick up some gifts from her and the children.

Still, the cake remained the top priority and, as she lowered the scalpel into the abdominal wall, Sherry informed the others she'd be departing prematurely in order to reach the bakery in time. "With my luck, all the good cakes will probably be gone," Sherry fretted as she pierced the epidermal layer and began reflecting the layers of skin to expose the underlying muscles.

Just then Ivan showed up, an hour late. "We did the abdomen, you have to do the scrotus," Sherry informed him. Ivan grimaced at the penalty exacted for tardiness. He explained his delinquency as a sleepless night antidoted by a mental health morning that stretched into an early-afternoon haircut.

Although drifting off had never been a problem for her, Sherry nonetheless commiserated with Ivan over his insomnia. "When I hit the pillow I'm gone and lately I've been sleeping like this." She folded hands over chest. "Put a lily in my hand and I'd look like I'm dead. Just like Lilly, come to think of it. Remember her? The way she used to sleep on *The Munsters*?"

Actually, no one remembered: For children who'd come of age during the Reagan administration, *Family Ties* was the sitcom point of reference for the rest of the table.

Sherry glanced at Udele, who, as she often did while someone else was cutting, looked like an animal of prey ready to pounce at the first sign of vulnerability. "Ready?" Sherry asked, unnecessarily: Udele already had scalpel in hand.

After trimming the skin from the bottom of the abdominal wall, Udele set the scissors down and calmly slipped a finger into the scrotum, extending it outward.

"That's how you do it? That's how you find a hernia? Just come at it from the other direction, stick your finger in the scrotum? I didn't know that," Sherry said.

"Me neither, never been there," Jen deadpanned.

"And then what happens, I guess, is when they cough the intestine drops through and you can feel the hernia through that opening? You mean the intestines actually break through that little opening at the top of the scrotum? Wow," said Sherry.

Eavesdropping, David Murphy turned from Table 25: "Not the most socially comfortable situation," he said, grinning.

Assigned to Lab C for the third and final unit were two adjunct itinerant instructors, Dr. Barry Grunwerg and Dr. Arvindkumar Suthar. Due to the ongoing faculty shortage, Vasan and Dr. John Siegel had recruited Grunwerg from NJMS's companion institution, the UMDNJ-Robert Wood Johnson School of Medicine in New Brunswick, where the mild-mannered dentist served as assistant coordinator in the gross anatomy curriculum.

Suthar spent his mornings teaching in an anatomy lab at the Albert Einstein School of Medicine in the Bronx, dashing across the George Washington Bridge to make it to Newark by the time the students reported to the NJMS lab at two o'clock. His work in other anatomy labs—Suthar also worked at New York University—made him despise all the more the OSHA-imposed ban on formaldehyde foisted upon Roger Faison. "They say [phenol] is safer, but the condition of that body down there"—he gestured toward a decomposing cadaver—"is much more dangerous to the students than formaldehyde."

A short man with a hook nose and impish personality, Suthar's trademark was the oversized forceps he used as a dissecting tool, a pointer and, when need be, a baton with which to animatedly punch the air to underscore a specific detail.

That afternoon, a wide-eyed Udele got her first look at Suthar's

legendary forceps when he thrust them at a cluster of grape-sized
knots appended along an abdominal nerve system.

"Know what these are?" he asked.

Udele looked at him blankly.

"Lymph nodes," Suthar said, answering his own question.

"And this is the inguinal ligament, where hernias reside, break
through." The instructor paused. Then said, "Honk-honk."

"Pardon me?" Sherry scrunched her face.

"You know, what's the honk-honk commercial?"

"Double A, M-C-O?"

"Right, think of the commercial, that's the mnemonic to remember
the key parts of the inguinal," said the instructor, initiating an
amphigory recitation . . . "areolar tissue . . . aponeurosis . . ."

Upon completion of a mnemonic understood by not a single member of his audience, Suthar reached in with his right hand and partially removed the intestines from the lower abdomen, piling the
organs atop one another until they resembled grotesque link
sausages. "How gross is that?" said Sherry. Udele cringed.

Totally unaffected, Suthar jabbed the forceps at a descending structure resembling a nerve. "This," he said grandly, "is the spermatic
cord. You can remember it, too, with a commercial: 'See Seaman's
First.'"

Once the groans subsided, Sherry posed a series of questions about
vasectomies, specifically the most effective means of performing the
procedure. "You can go through the abdomen, but the best way is
through the scrotum. That way you don't have to cut through too
many tissues. The best surgery is cut through the fewest number of
tissues," said Suthar.

Tentatively—Ivan much more reticently than the others—the four
lab partners took turns probing the scrotum, gingerly testing on the
cadaver the necessary but, as David Murphy had pointed out, socially
uncomfortable technique each would be called upon to perform thousands of times in the future.

Other tables approached the exercise with mirth, an opportunity
for double entendres and casual asides about comparative anatomy.
Table 26 went about their business sternly, without the usual repar-
tee, treating the patient with a gentle dignity, as though he were alive,

sentient. Whenever Table 27 visited—during the third unit Table 26, with a male cadaver, and Table 27, with a female cadaver, spent a good deal of time moving between each other's work stations to familiarize themselves with genitalic deviations—Jen, Udele, Sherry and Ivan made sure their classmates did the same.

While her lab partners poked and prodded, Sherry quietly slipped out, headed for the bakery, leaving Udele, Ivan and Jen as the sole student occupants of Lab C. Jen delightedly took note that they were not totally alone: Grunwerg was there, too, letting them know that for the remainder of the semester he'd remain in lab as late as they wished. Jen thought his presence a "good sign."

Udele perused the *Dissector*, studying the instructions for the day's final task. "You want to do it?" she asked peevishly, offering Ivan the scalpel. Her invitation politely but firmly declined, Udele, assisted by Grunwerg, split the scrotum in half, scooped out the spermatic cord and let it lay flat in her fingers for inspection.

Ivan and Jen's attention, however, was on the clock. "It's late, let's look at it tomorrow," Jen suggested, to unanimous consent.

The bakery was sold out of cakes when Sherry arrived, forcing her, unnecessarily as it turned out, to pick up a $13.95 ice cream cake at Carvel's—sensing Sherry wouldn't get out of Newark in time, the children's nanny had managed to squeeze into her schedule a trip to the local bakery.

"A nice cake, with candles, too," Sherry reported icily the next afternoon, picking up the syllabus and taking heed of the admonition: "Note: Orient yourself first, and PLEASE DO NOT DISSECT at this time."

Doing precisely as instructed, Sherry assiduously identified the diaphragm, liver, stomach, spleen, small intestines, large intestines, the transverse and descending colon, stomach, rectum and appendix. All structures were not present and accounted for.

"I can't find the spleen, where's the spleen?" Sherry asked loud enough to get Suthar's attention. The instructor came running across the room, oversized forceps in hand. "Look under the ascending colon and splenic flexor," he recommended.

"But why is it there?" Sherry wanted to know.

"Because that's where they put it." Suthar shrugged.

It soon became evident that Sherry would hold sway in this particular endeavor, a circumstance made possible by Udele's absence that
day, the first time she'd missed lab all semester. Given what Ivan and
Jen feared Sherry might find in the GI tract, they ceded her control
without protest.

Reaching under the liver, Sherry fulfilled her lab partners' every
qualm. "I think I found something here that isn't, uh, nice," she said,
brown matter dripping off her gloves.

"My God," said Jen.

"It's OK, really, I think it's hemorrhagic." On closer examination,
the substance appeared more red than brown.

"Well, I guess that makes it a little better, then," said Jen, unconvinced.

"So, where's the gallbladder? Isn't that supposed to be here somewhere?" Sherry muttered. Suthar reappeared, submerging the forceps beneath the liver. "It's supposed to be." He prodded aside the
liver. "But he doesn't have one."

Jen grimaced. "Gallbladder surgery, *too*? Poor guy."

Sherry dug right back in, hypnotically fondling the liver, then the
intestines. "You're having way too much fun," Jen informed her.

"I know, I'm disgusting myself. After twenty years of watching
them do this, now it's my turn. I always wanted to get my hands on
some guts," Sherry replied, extracting another handful of brown glop.

"Oh God, oh God, why is that so gross and why does she keep
going back for more? She loves the brown stuff. You do, don't you?
You love the brown stuff. Do you think it's fecal matter, maybe a hole
someplace?" Jen wanted to know.

Without thinking, Sherry sniffed her gloves. "I don't think it's
fecal." She wiped her hands on the paper towel, prompting Jen to
focus on another indeterminable substance, the evidence of which
could be seen through her lab partner's translucent surgical gloves.

"Are those *sparkles* on your nails?"

Sherry threw back her head and laughed. Glittering fingernails
immersed in entrails; it *was* incongruous. On Sunday, with Sherry
preparing for the gross anatomy exam to the exclusion of all else,
eight-year-old Sarah approached her mother, discount-store nail kit
in hand, with a proposition: "Mom, if you can't pay attention to me,

can I do your fingernails while you study?" Four days later, the glitter remained, a reminder—as if Sherry needed another after the previous evening's cake fiasco—of what was being sacrificed.

Happy to change the subject, Sherry returned her attention to the intestines spilling from the abdominal cavity. "I'm having trouble getting oriented, I think, maybe, because he's shorter," she said, visually calculating the distance between the head and pelvic region. Ivan, until then a quiet observer, modestly interceded, guiding his lab partners through the serried small intestines, including the difficult-to-discern duodenum, jejunum and ileum, with an expertise that clearly impressed Jen and, more important, Sherry. Ivan these days was doing his homework.

It being Sherry's day, her hands were back in the morass the moment Ivan completed the tutorial.

"You are the Unit Three queen," Jen told her.

"I just love the way this feels. I guess that makes me a little warped," Sherry confessed.

"You're going to be one of those doctors who asks every five minutes if the patient has had a bowel movement," said Jen.

"And the consistency of it," Sherry added, her demeanor suddenly becoming more serious with the realization that her self-indulgence, perverse as it might be, was short-lived. Once again, a physiology exam threatened to play havoc with the tables' priorities. "What are we going to do about tomorrow, physiology or anatomy? It's a big day tomorrow in anatomy, but we have the test . . . I'm not sure what we're supposed to do," she asked.

Jen, projecting ahead to a day when the dual demands of anatomy and physiology would undoubtedly seem barely aggravating, drew an analogy from her possible future in emergency medicine. "It's like being in the ER when someone comes in with V-Tach [cardiac arrest] while you're working on a broken arm. You make sure the broken arm is comfortable and then take care of the V-Tach. Right now, anatomy is the broken arm, phys is the V-Tach," she said, her voice trailing off with the appearance of Vasan, Varricchio right behind him pushing the cart holding the anatomy exams.

Vasan congratulated Lab C for contributing to a mean score, 82, four points above that on the first exam. Then came the rejoinder:

"Good grades were essential on this one because bad days lie ahead," he said, referring to the onslaught of exams in the offing.

Content to hang back while the rest of the room flocked to Varricchio to retrieve their exams, Table 26 amused themselves with an examination of the pancreas and were surprised to find the cadaver's appendix intact.

"Weird; it seems like it's the one thing they haven't taken out of him," Jen volunteered.

Finally, Ivan broke ranks. Handed his exam, he peeled back the first page with trepidation, took a deep breath and looked at the number scrawled within: 87. "Excellent," he said, breaking into a wide grin.

A second later, Jen and Sherry traipsed over to receive their grades. As expected, Jen again achieved high honors; Sherry, as expected, did not. Unlike Ivan, they responded impassively.

Sublimating her enthrallment with the contents of the abdominal cavity, Sherry spent nearly every waking minute over the next ninety-six hours enmeshed in physiology. Late Sunday afternoon, the wet snow that had blanketed the state earlier in the day knocked out an electric transformer on Sherry's block, plunging the family into darkness and Sherry into a state of panic. Her attempt to study by candlelight thwarted by four restless children unexpectedly thrust into a VCR-less world, she "completely lost it." "My life is a mess; this was just what I needed," Sherry said the next day, relating the story.

Stepping in before chaos turned catastrophic, Jerome packed the kids off to his mother's house, where the electricity and VCR were still in working order, and then deposited Sherry in his office at police headquarters. To minimize further distractions, he turned off all the scanners and quietly shut the door, leaving her to contemplate physiology as well as the mug shots decorating the walls of her husband's place of business.

Following the exam, Jen and Sherry had the table all to themselves. Ivan, exhausted, had gone home to take a nap; Udele, for the third time in under a week, was inexplicably a no-show, causing Jen, who tended to take things personally, to wonder aloud if they'd somehow offended their lab partner. "Can't think of what it was," said Sherry, intent on rekindling her love affair with the abdomen.

Tracing the digestive system downward from the stomach, she discovered that Suthar's forceps had the previous week betrayed him: A gallbladder indeed existed, although, Sherry had to admit, it was pretty much flattened beyond recognition. To Table 26, the presence of the gallbladder meant that, unlike the thoracic cavity, their cadaver's abdomen wouldn't become a destination for the rest of the class. Furthermore, not only did it appear that the cadaver lacked any abdominal surgery, Table 26 never encountered any of the marble-sized gallstones that many of the others, with unabashed glee, were digging out of their bodies.

Locating the pancreas, Sherry noted that its position and function explained why pancreatic cancer spreads so quickly to other organs.

Next she lifted the stomach out of the abdominal cavity and demonstrated its elasticity, stretching it wide to show Jen what it might have looked like when the donor complained of being full. A bisection of the organ laid bare the rugae, gastric folds within the lining, the keys igniting the digestive process.

Nearly identical to the census undertaken the first day of Unit III, Sherry's tour of the abdomenal cavity revealed a palate of well-articulated organs, positioned for maximum effectiveness, the ideal blueprint to maximize Vasan's emphasis on embryonic development.

The abdomen's major flaw was the extraneous contents that, on occasion, even brought Sherry pause. "Is that . . . poop?" she stammered, watching Grunwerg perforate the duodenum.

"Oh, come on, Sherry," Jen retorted, "this is your unit, you should be liking this."

The unit that so captivated Sherry evidently held no interest for one of her lab partners. And when Udele didn't show up again the next afternoon, Jen ratcheted up the previously expressed concern that the rest of the table might have done something to drive her away.

"I know she's here because I saw her before," Jen told her lab partner. "Maybe she doesn't like us anymore." She paused. "That would hurt my feelings."

"Mine, too," Sherry said, following up on the televised prelab lecture by flipping the *Dissector* to page seventy-three, the section dedicated to the study and dissection of the kidneys. *Grant's* stipulated that the kidneys be accessed by cutting from the front of the cadaver through the protective covering known as the parietal peritoneal of the posterior abdominal wall, a course of action that spared everyone, for the moment, anyway, the trauma of another flip.

Sherry lifted the intestines, pancreas and liver, located the parietal peritoneal and excised the membrane with scissors. "I believe this is the kidney, but I could be wrong," she murmured, staring at the glut of tissue inside the retroperitoneal cavity.

Jen picked up the *Dissector.* "It says here, we're supposed to be looking at the right kidney, dissecting the left. Maybe we should go for the left."

Intent on finishing what she'd begun, Sherry continued to reflect the tissue around the right kidney. Jen observed for a minute before shifting to the other side of the cadaver to excise the parietal peritoneal sheathing the left kidney, a maneuver that left her disoriented and confused.

Sherry, faring no better in locating the right kidney, consulted the irreparably stained *Netter's Atlas* for an answer. The trick, she told Jen, was tracing the path of three key arteries leading to the kidneys.

"Here's one," Jen said, pointing to the mesenteric artery. "I think, over there, you need to find the aorta and the testicular artery. If you find the testicular and follow it up, it's supposed to put you right there. The book says the right [kidney] is higher than this one."

Jen let her hands drop. "I don't like all this grabbing and pulling when I don't know what I'm doing," she said dispiritedly as Suthar materialized, most opportunely.

"We're lost and it's just not a good place to get lost. I think we need to find the testicular, but I can't find it anywhere," Sherry told him.

Suthar curled his upper lip in disgust. The problem, he assured Jen and Sherry, had nothing to do with their skills as anatomists. "It's the bodies," he complained, placing the forceps into full orchestral conducting mode. "Look at them. Disgusting. They aren't well preserved, that's why it's difficult to find it."

The instructor scraped the tissue just below the right kidney with a hemostat. Thirty seconds later, a glistening artery, the testicular, revealed itself. With a gloved hand, Suthar followed its path upward, stopping when he reached a recessed sac. "I believe that is what you were looking for," he said, pointing to an oblong object the shape of a computer mouse.

Sherry brightened. "Did you ever notice the minute they're [the instructors] here, we suddenly get real smart?" she asked, removing the right kidney from the sac, extracting along with it "more brown gunk."

Having located the right kidney, Sherry leaned across the table to ascertain why Jen was having so much difficulty on the left side when, out of nowhere, Udele appeared, slipping into the spot vacated by Suthar.

Never one to suppress emotion, Jen looked as though she might lock her in a bear hug. "Where have you been?" she blurted out. "We were worried about you. We thought you didn't like us anymore. You still like us, don't you?"

Udele blushed, her smile creeping ear to ear. "No, no," she said, pleased but nonplussed by the attention. "I mean yeah, of course I still like you," blaming her belated appearance on the hour she'd spent huddled with Vasan, protesting what she felt had been an ambiguous question on the practical.

"Speak of the devil," said Sherry. The course coordinator had entered Lab C and was heading their way. Aware of the demands on Vasan's time, Jen wasted not a second on niceties.

Gesturing at the right kidney, she informed him: "We found that one, but we're having a lot of trouble here." Vasan dug intently into the tissue surrounding the left peritoneal cavity, his brow furrowed in concentration. "Hmmm, there seems to be a condition here. Let's see if he had a renal problem," he said.

"Oooh, a renal problem, *that* would be interesting," said Sherry; she shared Vasan's enthusiasm for surgical legacies. To Sherry, the cognitive ability to understand and repair the multitudinous components within the human body was a miracle equal to its creation, however it might have come about.

Often, Vasan explained, hypertension can lead to arterial swelling, manifesting itself in kidney failure. "If the testicular vein swells up, like a hernia, it could be a real problem. Let's follow it along," he suggested, tugging at a structure that could have been a nerve or an artery.

"What's that?" Ivan asked.

"Don't know yet. I have to see the structure, where it originated; I want to look a little further before I say," said Vasan.

"I have a question," Ivan said deliberately. "If we see a structure during the practical and we can't name it, can we say, 'I want to look further'?"

Vasan peered over his glasses. "Nice try," he said, returning to the mystery structure. "This is it, what you need: the testicular. Follow the vein up, here. And to open the kidney you need to go right . . . there." Vasan stabbed a finger at the membrane covering a sac identical to the one from which the right kidney had been extracted. "Look for the super renal vein, the urethra, follow it up. The left kidney will be right there," Vasan instructed, leaving Table 26 to their own devices.

Poking around the left peritoneal cavity, Jen located the super renal vein and examined again the testicular artery. "This is weird. There seems to be some abnormality over here. Something doesn't add up," she said quietly enough that no one else—such was the rest of the table's fixation on the exposed right kidney—heard her.

As the forceps inched to within a centimeter of the sheath obscuring the left kidney, Jen suddenly diverted her attention to the progress being made by Sherry opposite her. "This side of the body is not so nice," she declared.

Sherry nodded her head in assent. "Let's just do this one, it's already out," she suggested, cleanly bisecting the right kidney, not bothering to gain a consensus.

"Why not? They're both the same," Jen agreed, leaning across to more closely scrutinize the kidney laid open by Sherry, thus aborting the table's examination of the content within the left peritoneal sac.

CHAPTER 15

Ivan Gonzalez was on a roll, distinguishing the structures inside the cadaver instantaneously, sometimes before the forceps in Priya Singh's hand caressed the tissue she asked to be identified. "Good! Great!" Priya exclaimed with each correct answer. Brazen in his grasp of the subject matter, Ivan resembled a prizefighter, rocking back and forth on the balls of his feet, anticipating the onslaught of questions thrown at him. It was a side Ivan rarely showed his lab partners. Nor were they seeing it now, with the hour fast approaching eleven o'clock on a Saturday night.

In slightly more than thirty-six hours the fourteen-week sprint through human anatomy—an amount of time neither faculty nor students thought sufficient—would be over. Two days hence, when Ivan awoke, there'd be no more lab, no more first-year smell, no more daily contact with Sherry, Udele and Jen. Ivan wasn't certain how he felt. Relieved, to be sure. But also slightly melancholy that an endeavor which had consumed so much emotional and intellectual currency, both prior to coming to medical school and especially over the past four months, would soon exist only in the past tense.

He'd made it, of that Ivan hadn't a doubt. Failure, which loomed before him barely a month before, was no longer plausible. At low ebb after the first anatomy exam and second physiology test, he'd sucked it up and done what was necessary by jettisoning every aspect of his life detrimental to attaining a degree in medicine. The commitment, he admitted, further eclipsed the character trait that had once defined him as a person others turned to in moments of need. Superficially friendly and outgoing, Ivan regretted that, inwardly, the reallocation of his priorities had metastasized into an aloofness over which he had no control. Still, while he didn't particularly care for the person his ambition had created, validation came from knowing he would not have to repeat gross anatomy during the next academic year or, worst-case scenario, return to Miami humbled, defeated, never to be a physician.

Arduous though the semester had been, there were parts Ivan was going to miss, principally the man on Table 26, a person about whom Ivan knew nothing and everything. He truly liked the guy, an unrequited rapport that often demonstrated itself in unconventional ways. Like the Saturday night before the final exam when he dragged Priya over to show off the cadaver's "nice and round" prostate.

Ivan had arrived at the lab at 10:30 after spending the morning and afternoon at his apartment consuming *Snell's Clinical Anatomy* and the early evening hours closeted in a study room scrawling crude renditions of circulatory and nervous systems on the wallboard. In lab, he hooked up with Priya Singh, the young Indian woman with sparkling brown eyes and easy smile who, had the class voted on such things, would have won Miss Congeniality hands down.

Together, Priya and Ivan hopscotched from room to room, directed by fellow nocturnal travelers to specimens of specific structures likely to be tagged for the exam. In Lab A, they examined an exemplary medial femoral reflex artery in the leg; Lab D boasted a prominent umbilical vein exhibiting itself in the abdomen; a table in Lab B distinguished itself by hosting a cadaver with external iliac nodes inside the lower abdomen.

It was a crapshoot, plain and simple, matching the students' propensity to guess which structures would be tagged with what would actually occur once the faculty filtered in Monday morning to play Santa Claus one last time. The scene, three dozen young people tearing around a room filled with corpses in the wee hours of the morning, was the penultimate chapter in an episode the participants would undoubtedly come to view as primarily enlightening and almost equally outlandish.

"Look around here. Can you believe these are human bodies? It's hard to imagine," said Priya, taking into account the transmogrified accumulation of muscle, bones and tethered skin sprawled atop each table. Priya and Ivan's intralab sojourn brought them back to Table 26, where, with a smattering of pride, Ivan showed her the cadaver's lateral fibular collateral, the knee. Priya was impressed. Though equally fond of the cadaver at Table 27, she recognized that "Elizabeth" left much to be desired when it came to musculature and bone consistency.

Prayerlike, Ivan folded his hands and recited the structures inside the pelvis; Priya filled in the blank each time he hesitated. "Listen to her, she's a book. Somebody should marry her, she's got the stuff to be a neurosurgeon," Ivan marveled.

"A neurosurgeon of the pelvis? That's an interesting concept," said Mohan Madhusudanan, a student who'd joined up with them during the earlier translab pilgrimage.

Summoned to Lab B, where rumor had it there existed "the best prostate in lab," Mohan and Priya abandoned their friend, leaving him to recount alone the edifices near and around the kidneys.

Ivan found the left phrenal vein and began recalibrating the path taken by Jen three weeks before. At the left parietal peritoneal he hesitated. With furrowed brow, he poked at the tissue with a hemostat. "Where's the kidney?" Ivan muttered to himself. He stood at attention, eyes shut, his mind replaying the events of the afternoon when they worked on the kidney. . . . Jen was on the cadaver's left. . . . Sherry on the right. . . . Ivan's eyes snapped open. They hadn't dissected the left kidney! Reaching across the table, he plucked the right kidney from its sheath and, with his fingers, examined Sherry's dissection.

In a unit encompassing the abdomen to the toe, Ivan could be forgiven a memory lapse. All in all, GA's third segment had been a blur, a five-week stretch broken in the penultimate week by spring break. Wanting nothing more than to go to Miami—Ivan missed his backyard, the warm weather and even the driveway that spared him from trying to find a parking place every night—he had remained in Newark, exiled by the physiology exam scheduled for the day classes resumed.

Compared with the intricacies endured just prior to break when they'd plumbed the abdomen, pelvic area and genitalia, the last week of gross anatomy lab had been a snap. The dissection of legs and feet, structurally analogous to the upper appendages, mimicked the procedures performed two months before when they'd worked with the arms and hands. Going through the motions, everyone but Jen was completely bored by the process. What the abdomen had been to Sherry, the legs were to Jen: a source of relentless fascination. Jen loved the way the tendons fused with muscle and bone and cartilage, a symmetry she categorized as "funky."

Funky was the last adjective Sherry would have conjured to describe a section she considered tedious at best. In an attempt to make things a little more stimulating, she began searching for the arteries and vein absent in the legs after being transplanted into the thoracic cavity during what she figured had been Number 3426's first bypass surgery. When that proved equally soporific, Sherry turned to cause and effect.

"I'm a puppeteer," she declared, retracting the flexor digiti minimi brevis manus, allowing the cadaver to perform a feat in death that never could have been accomplished in life—wiggling the little toe independent of the others.

Mostly, Sherry shuffled somnolently through the fourteenth week of lab, her focus diverted from the verities of the leg to the overarching priority of getting forever out of there.

Exhausted by the unrelenting barrage of exams—midway through the semester psychology had been added to the schedule—those at Table 26 were experiencing tension during the days leading up to spring break that had become palpable. Udele displayed only flashes of the earlier ardor she'd had for dissecting. Ivan transformed himself into the stealth member of the team, as likely to be found at another table as his own. The gauge of Jen's waning enthusiasm, even during the dissection of the leg, was the near total elimination of the word *cool* from her vernacular. And Sherry, vowing to no longer wallow in self-pity, nonetheless still coveted the flexibility of those in class who didn't have the responsibility of looking after a husband and four kids—in other words, everyone but her.

Still, at least Table 26 continued to show up every afternoon. Which is more than could be said for many of their classmates. "I can't believe they can get away with blowing this off. If I don't see this stuff with my own eyes and touch it with my own hands, then I don't learn it," said Sherry, unaware that many of the daytime absentees were returning to the lab at night.

Coming in after the adherents to the daytime schedule had departed provided the so-called night crew a measure of freedom from the faculty and, for some, emancipation from lab partners whose company had worn thin. By the midpoint of the semester, divorces among lab partners were commonplace. One estrangement took place at Table 37, where the metabolism of the cadaver and the relationship among the four students assigned to dissect it deteriorated at an almost equal rate.

Ann Waldman, who'd related to Jen the contents of her nightmares the first week of class, was the first to defect, finding a more benevolent home at Table 39, where Bradd Millian, Gayatri Rao and Elizabeth Garcia held court. A week later, the cadaver reeked so horribly the remainder of the room demanded that Table 37 shut down operations.

Besides mitigating whatever tensions existed between lab partners, working at night allowed a level of independence that instigated a minor controversy midway through Unit III.

The stories about anatomy students run amok are so prevalent in medical schools that inevitably they've seeped into the public domain. In the most common telling, human skulls are utilized in impromptu anatomy lab football scrimmages or severed human arms become rapiers in lighthearted lunge and parry between lab partners. It speaks volumes that whenever the callous mistreatment of cadavers is chronicled, the incident always seems to have occurred at a school other than the one where the story is being told. There are two reasons for this: first, an awareness among anatomy students that such behavior would never be tolerated at *their* school, and second, the tales are, at face value, urban myths, or the medical school equivalent thereof.

With over twenty-five years spent in anatomy labs, Nagaswami Vasan's closest encounter with anything even approximating desecration came one Halloween when a student outfitted a cadaver in a three-piece suit. The student was allowed to remain in school, but only after coming within a hair's breadth of expulsion. At NJMS, treating the men and women who donated their bodies with dignity was a mantra established at the opening lecture when Dr. John H. Siegel told the Class of 2002 before they ever picked up a scalpel, "Have reverence for the cadaver. For it is, in fact, a person who has made a sacrifice for your education."

While the need for reverence went unquestioned, everyone

involved recognized, if only tacitly, that the human psychological condition demanded levity play a role in what, bottom line, was a particularly heinous experience, the deconstruction of human bodies. No one could deny that humor took the edge off, but students often parted over interpretations of what, precisely, was and was not funny. Nearly every day, in one way or another, someone in lab managed to prove true the old adage about there being a limit to good taste but no limit to bad taste. As Dr. Todd Olsen, the course coordinator for the anatomy curriculum at the Albert Einstein School of Medicine, pointed out, the human proclivity to diffuse tension in anatomy labs with humor "often brings out the worst in people because one person's joke is the source of another person's anxiety."

It wasn't clear whether what occurred at NJMS as the semester wound down was a product of curiosity, passive-aggressive behavior, misdirected humor or, as Sherry noted disdainfully after learning the details, "a lack of maturity." Jen ascribed it to all three and wasn't sure what bothered her the most, the fact that a classmate had returned to lab after hours to circumcise her cadaver, a procedure neither authorized nor sanctioned by the anatomy department, or all the attention the student was receiving from fellow students as a result.

"That's just not right, he didn't ask for that to be done, she did that without his permission," said Jen, her face flushed in anger. By no means a prude, she'd been an active participant all semester in the good-natured bantering during which the cadaver sometimes acted as foil. Humor was one thing; violating a cadaver's dignity and, by extension, the inherent trust he'd endowed to those to whom he'd donated his body was quite another. Just as Jen wondered if she should confront the perpetrator of the circumcision, Sherry quelled her lab partner with, of all things, a dose of med student humor.

"Well, at least he didn't feel anything. That's better than they usually have it during a circumcision. I think they should be anesthetized. Of course, I think everyone should be anesthetized; it's good for business," said the nurse-anesthetist.

"OK, let's do it," said Jen, returning the focus to another unpleasantry: a flip. Grabbing hold of Number 3426's left foot, Jen directed her lab partners to assume their stations. "Ready? OK. One . . . two . . .

three aaand *lift.*" Afterward, Ivan elevated the pelvis to allow Sherry access for a woodblock to be shoved underneath, turning the body into an inverse V, the buttocks elevated a foot above the table.

Surveying the tableau, Jen enthused, "This really makes me feel like a doctor. I mean, there's no other situation in which you'd find yourself in this position."

"I guess that depends on your social life," said Sherry, not missing a beat.

Although the day's assignment, dissection of the anal triangle, had been presented in prelab as a "clinically important area," the source of the nervous buzz filling the room had more to do with expectation of encountering matter of a more extraneous nature—hemorrhoids, bits of toilet paper and fecal matter. Sherry did a quick inspection, reporting, "It's not that bad."

Unwilling to take a chance, Jen discreetly dabbed the region with a damp paper towel. "I guess I really don't need to do this," she explained, "but a friend of mine at Einstein, a third year, told me that the one thing you learned from doing this is that men don't know how to clean themselves down there."

For the edification of the others—this seemed like something that needed to be shared—Jen began to read aloud from page eighty-six of the *Dissector,* pausing when she got to the part about cutting through the fibers near the surface of the sacrum and coccyx to sigh, "this is sounding gross already."

Having exhausted what had once seemed an inexhaustible predilection to dissect, Udele stood passively to the side, posing not the slightest impediment to Ivan when he seized the scalpel. "Do you think we should practice the rectal?" Jen asked at the moment her lab partner was poised to make the incision.

"Be my guest," Ivan said, with dramatic sweep of his hand.

"I think we're supposed to; that's what it says in the *Dissector,* anyway. I asked my husband last night if I could practice on him," Sherry said.

"No way." Jen laughed.

"I did. He wouldn't let me; he's just no fun."

Jen considered the timing, wondering if it might be better to wait until after they'd dissected: "I don't know if we should. I don't feel right about it. I want to check his prostate, but sticking my finger up there doesn't seem right. I mean, what's the difference between doing that and what she [the student who performed the circumcision] did?" said Jen, nodding for Ivan to proceed.

Ivan's perfunctory incision transformed what a few moments before might have been an invasive procedure into an anatomical examination, the only encouragement Jen required to perform her first rectal exam.

The procedure, indeed the entire dissection of the anal triangle, didn't live up to its horrendous billing. In fact, Jen found it rather mundane, just another installment in the compendium of parts she'd been memorizing since January, different from the others due mainly to the fascination and revulsion attached by society to the function that, for those in good health, occurs each day in that sector of the body. If anything, Jen thought the experience grounding, a reminder that despite the high-minded rhetoric about the superiority of human intelligence, ultimately "we are all animals" born of and forever possessing a primal urge to eat, urinate and defecate.

Overall, the anal triangle proved elucidative, an opportunity to understand a spectrum of clinical objectives that ranged from performing a sigmoidoscopy in a manner minimizing "unnecessary discomfort" to determining which nerves cause the involuntary contraction of the rectal muscles during death by asphyxiation, specifically hanging.

Having triumphed over their anticipated aversion to the region, access became the table's primary obstacle: They just couldn't get a decent look at what they needed to see. Even worse, they were receiving mixed messages on the propriety of proceeding with the solution to their dilemma.

The problem began with Vasan's ambiguously worded directive that the students limit access to the pelvis with a longitudinal bisection—in effect, splitting the pelvis down the middle—and not a full pelvic hemisection, which would also separate the lower half of the body from the top. Half the faculty interpreted the injunction to mean Vasan had given each instructor the discretion to decide whether to proceed with a full hemisection; the other half swore Vasan had issued an outright ban of the procedure.

Until that juncture, the instructors had done a fairly good job of glossing over philosophical differences about the direction and cadence of the course, reserving criticism of one another, and especially Vasan, to pointed asides outside the earshot of the class.

Take eight faculty egos, each intellectually capable of grasping the complexities of human anatomy, throw those egos together for four months and there was bound to be a point of combustion. The full hemisection was that intersection. In their open defiance of Vasan, the Class of 2002 got caught in the middle, none more so than those assigned to Lab C, where one instructor, Suthar, turned a blind eye to renegade tables that went ahead with the procedure while the other instructor, Grunwerg, the outsider recruited from New Brunswick, toed the company line.

Suthar at least had tradition on his side since, in years past, the technique had been part and parcel of the pelvic dissection. But tradition made it no less difficult for the students, who couldn't determine which course of action they were expected to take.

Udele, Jen, Ivan and Sherry needed only visit their friends at Table 28 to see the advantage of the full hemisection: an unobstructed view of the structures within the pelvic girdle. "Look at that," Ivan enthused, his élan undiminished by the observation of the person who had performed the procedure—"By far, the most disgusting thing we've done yet"—or his own role as the designated handsaw operator should Table 26 decide to buck convention.

Table 26, though, was not of a mind to act as renegades. Not yet, anyway. Waiting for a signal from either Suthar or Grunwerg about which way to proceed (Vasan had already weighed in, pointedly advising them not to follow the lead of Table 28), Table 26 considered their options. "Want to do the penis as long as we're here?" Udele asked.

"Might as well," Jen shrugged, receiving from Udele the reward for her acquiescence, the scalpel.

"How are we supposed to cut? Top to bottom?"

"It says make a midline incision," advised Sherry, consulting the *Dissector.*

"Where's the muscle, find the muscle," Udele interjected. Elsewhere, pockets of nervous laughter indicated others were at an identical stage of the dissection. Jen refrained from cracking wise, setting the tone for the rest of the table. Taciturnly moving the penis out of the realm of mirth, mystery, joy and consternation, she made sure her lab partners viewed the appendage as part of an amalgam that, for their purpose, needed understanding only from the perspective of anatomical purpose.

Understandably, the enterprise in which the class now found themselves engaged had undertones that summoned the best in some, the worst in others. "Intellectually, it is difficult to appreciate the genitalia, internal and external, and when you need to learn this appreciation in mixed company it is a very powerful experience," said Olsen, the course coordinator at Albert Einstein.

At one end of the spectrum, the anxiety cloaked itself in jokes of the Lorena Bobbitt genre or even more personal levels. "Man, I'm sure glad I got laid this morning," said a male student, not too loudly, as he examined what remained of a dissected vagina.

At the other end, Leah Schreiber lacked the wherewithal to sink to smarmy, below-the-belt humor. Observing the dissection of the vagina, Leah fervently wished "we could go back and do the head again. . . . I don't know how to explain it but some things still make me feel uncomfortable, like peering into someone's private area, even if they are dead.

"Does it make you uncomfortable?" she asked Christine Ortiz, her lab partner.

"It did the first time I had to check an episiotomy," the former nurse admitted. "But then I got used to it. This, dissecting it, doesn't bother me."

"I don't know why it bothers me," Leah continued. "Maybe because I've seen her at an angle that I've never seen myself. I think this brings it all back, the humanity and everything. I mean, we now know why she donated her body to science—it's because she had everything in the world wrong with her: an umbilical hernia, pacemaker, cancer. She probably has so many experiences with medical professionals that she figured it didn't matter if she had some more after she died.

"Most of the time I don't think about it until you're doing something like we're doing today. Then you're really aware of her humanity. When you're in the abdomen or even the face or arms or hands you can pull back. You're dealing with organs and muscles and nerves and arteries and veins. It's easy when you're inside. When you're outside, when you're here, it's not. You feel embarrassed but you don't know why, but I think it's because you're remembering that this is a human being."

Six feet away from where Leah obsessed over issues extraneous to anatomy, Table 40 was consumed with a more technical matter, a penile implant. As Vasan and the other instructors had witnessed countless times, the discovery of the procedure proved irresistible to the entire class.

"Oh, that's terrible," said one student, inching her way closer for a better look.

"Terrible? What's terrible about it?" another wanted to know. "He had a problem. They fixed it. You're saying that's terrible?"

"I just can't imagine the trauma he must have gone through," said the first student.

"Ah, the pre-Viagra days," said Leslie Pooser, gaining the approval of Dr. Michael Rose, the instructor overseeing Lab D during the third unit. Using Leslie's observation as a starting point, Rose noted that sexual dysfunction becomes so defining that it becomes motivation for resolving the problem. With that, Rose purposefully began moving the onlookers back to their respective stations to allow Table 40 the opportunity to determine what had caused the man to seek the implant. Begin by reexamining the prostate, he recommended.

Their brief excursion to Lab D to see the implant did nothing to unlock Table 26's own confusion over the proper method of dissecting a human penis.

"Muscle? Did you say muscle? Does it say which way we're supposed to cut the muscle?" Jen asked Sherry and Udele. "The *Dissector* gives different directions than the prelab. Damn. Why can't they get together? The book says one thing and we get four different opinions from [the instructors]. It's just like with the hemisection."

"Start like this." Suthar, again, as was his way, appeared as if an apparition. Gently, the instructor peeled the skin off the appendage, specifying the pertinent landmarks as he went along.

When he finished, Sherry probed the urethra with a hemostat.

"Uh, couldn't you use something a little more gentle?" asked David Murphy, poking his head over from Table 25.

Sherry, perturbed, snapped back, "What's wrong? We do catheterizations all the time."

"Yeah, but those are with rubber tubes," Murph pointed out.

"But he's dead!"

"So, you use a hemostat?" Murph smiled with mock incredulity.

For Jen, the trip to Lab D had been illustrative on another front: She was now able to report that every table in that room had performed a hemisection. At Bradd Millian's table, number 39, the procedure had been performed minus the assistance of the latest addition to the team: Ann Waldman. ("I know now what my limits are and that, for sure, was beyond my limit," she said.)

Jen's revelation provided the impetus for the table to stop equivocating. "Let's go for the whole thing," she said, sending Ivan to the supply room to pick up a handsaw. Moments later, following Sherry's verbal instructions, he began bisecting vertically.

"That's the last vestige of his manhood," he murmured as the saw broke through the pubic arch.

"Is that what makes you a man?" Sherry wanted to know. Ivan ignored her.

Taking a look at Table 28, which had separated the left leg at the hip socket, Jen urged Ivan to plunge forward, turning the longitudinal procedure into a full hemisection. Following the directive, Ivan studied the pelvis, searching for landmarks. He didn't notice that Vasan had appeared, none too pleased.

"Stop right there, stop with the vertical cut. See what you have to see but don't do the full hemisection," Vasan ordered, strolling away before anyone could muster an argument.

Jen, disappointed, looked enviously toward Table 28, which, by disuniting the left leg of the cadaver at the hip socket, had a visual access to pelvic structures unavailable to her table. "I just wish they'd make up their minds," she said with resignation before heading to Table 28, the remainder of her team trailing, to see what needed to be seen.

Three weeks later, the furor over the hemisection long relegated to ancient history by the countless other mini-crises visited daily upon their lives as medical students, Ivan stood again at Table 28.

It was 2:30 A.M. Sunday and Ivan, drawing on a clinical question posed in *Snell* about the correlation between incontinence and inadvertent destruction of the puborectalis muscle during surgery, was once more availing himself of Table 28's fully hemisectioned pelvis.

With Priya fading fast, Ivan showed no signs of slowing down, enervated by the realization that his grasp of the material was better than he ever believed possible. The trick now was to carry the momentum into Monday. "I know my shit, I just don't have the confidence that [Priya] has," he said as Priya, the mischief in her sparkling brown eyes depleted by the late hour, called it a night.

Priya's departure dwindled the attendance in the lab to four. Hunched over *Snell's Clinical Anatomy, Netter's Atlas* and *Grant's Dissector,* a couple dozen more members of the Class of 2002 were scattered among individual study rooms, the cell and tissue biology labs, corridor sofas and chairs. One student fortified himself with an eight-ounce bag of kosher teriyaki beef jerky. Given the general texture of embalmed human skin exposed for fourteen weeks to scalpels and stale laboratory air, his classmates thought the food choice a tad disturbing.

The student defended his choice of sustenance by noting that only twenty of the seventy calories in an average serving were derived from fat. There must have been a caffeine element as well. For, with the exception of one other student, he alone in the cell and tissue biology lab had yet to succumb to the fatigue that had caused the other nocturnal warriors to pitch face-forward onto their textbooks.

At 3:30, Ivan lost the battle as well. The learning over and with the time remaining dedicated strictly to review, Ivan headed home for what he hoped would be one last decent night of sleep before the exam that would, at last, bring down the curtain.

CHAPTER 16

―――――――――●―――――――――

For the first time in weeks, they were all together, gathered as one, even the overwrought student who'd erupted into tears before the start of the previous day's practical exam. Vasan had quickly ushered the distraught woman out of lab and into the corridor, motioning for Paolo Varricchio to start the exam without her.

Shaken by what they'd just witnessed—Vasan hadn't reacted fast enough to prevent the rest of the students from observing their classmate's distress—a pall settled over the lab during the interminable wait for the horn to sound. Absent the furtive buzz of edgy humor that usually preceded a practical, the remaining students struggled to avoid eye contact, afraid it might betray the underlying fact that all present were, in one form or another, as fragile as the classmate whose implosion had, by all indications, just destroyed her every accomplishment along with her grandest dream.

What they didn't know until the following day was that, after they'd repaired to their apartments and homes for the obligatory postexam nap, Vasan brought the distraught student back to the lab where, alone, she took (and passed) the practical. Having already sent the rest of the faculty home, Vasan kept a discreet but watchful distance in his capacity as timekeeper and proctor; on the occasion of the final practical and written exam, Miller Time arrived way behind schedule for the course coordinator.

As much as they wished it were so, the final written and practical exam for Medical Gross and Developmental Anatomy was not the last time the class would be called upon to summon their expertise on the anatomical compendium. The denouement would take place two weeks later with the shelf examination administered by the National Board of Medical Examiners. The test would not only determine how NJMS students fared when compared with students across the country but also serve as a barometer for the program that had taught them. Despite the prestige riding on the numbers, neither faculty nor student body were predisposed to sweat the details since tougher questions had already been asked, in one form or another, on the departmental exams. Furthermore, the chance of getting ambushed by esoteric inquiries was greatly diminished by the course's primary textbook, Snell's *Clinical Anatomy for Medical Students*, which helpfully tendered sample shelf exam questions at the conclusion of every chapter.

Weighing far heavier upon them once the class had shaken loose the indigenous postanatomy test stupor was the ordeal of two physiology exams—the sixth biweekly benchmark test coming up on Friday followed by the final on Monday—bookending the weekend. Before the students submerged themselves in physiology, the department of anatomy mandated a diversion, an hour set aside in the frenetic calendar for a most unlikely medical school pursuit, personal reflection.

In the annals of medical education, holding a memorial service at the conclusion of gross anatomy to honor the men and women whose bodies nurtured the quest for knowledge is a phenomenon of recent vintage. Considered contrary to the detached pragmatism most programs believed essential to the effective practice of medicine, until twenty years ago only a handful of medical schools indulged in the practice. Today, the schools that haven't integrated the service into the curriculum give their students the option of holding a ceremony; in nearly every case the students exercise that option. Uniformly secular, the rituals go out of their way to avoid invoking a higher authority. Many schools invite families of the donors to attend the service, elevating the drama by introducing the next of kin to the students assigned to the dissection table where their loved one last rested.

The New Jersey Medical School, adhering to its policy of keeping the identities of the donors confidential, limits participation to students and teaching faculty. The exception, every year, is Essie Feldman, the ebullient secretary in the NJMS Department of Anatomy, Cell Biology and Injury Science, the only person in the building to know those being honored in terms other than the particulars of their respective anatomies.

The memorial service closed a circle begun in January when the tables were opened to reveal faces that once beamed in joy and contorted in sorrow, arms that enfolded husbands and wives, children and grandchildren, chests containing hearts that quickened with pride and also fear, and legs that, taken together, stepped on every corner of the globe. Forty-five tables, forty-five human beings, all of whom, when of sound mind, had, for the benefit of generations living and still unborn, agreed to the annihilation of every internal and external physical characteristic that ever defined them as such.

The deconstruction of those characteristics demanded dispassion. Not total indifference, but a dispassion streaked with respect and dignity, a Faustian bargain that provided the key to emotional survival. The students' purpose was to learn, not empathize; the secret to accomplishing that goal was to put aside the repercussion of the first encounter and to somehow teach themselves to forget that the faces had crinkled with joy, that the arms had embraced another.

Necessity marginalized the human beings into bodies, vehicles with glitches major and minor that became the underpinning in the quest of 181 men and women to heal well into the twenty-first century.

As per the signed agreements stashed in Essie Feldman's file cabinets, the students did precisely what the donors understood would be done. And though their bodies were summarily destroyed, the humanity of the men and women who'd donated those bodies, through the spirit of the gift they'd passed along, persevered.

Hence the memorial service, an opportunity to close the circle, a chance for students and faculty to say thank you, and good-bye.

For Vasan, the ceremony had always been a deeply personal experience. Just as he went out of his way to know every student in the class, so, too, was he familiar with the bodies in their care. Not only did the knowledge come in handy when identifying the best possible structures to tag for the practical exams, but Vasan also believed it key to the teaching dynamic: Inevitably, the peculiarities uncovered in lab one year were integrated into the lecture cycle of the next.

Vasan gave the selection of the students who would speak at the ceremony careful consideration. Reuven Bromberg had been one of the first approached. At that juncture, Reuven was considered a leader, a student unafraid to speak up on behalf of his classmates over injustices, perceived and otherwise.

Then came the hemisection of the head, a procedure Reuven continued to insist crossed the line between necessary academic behavior and desecration. That day marked the last many of his classmates saw Reuven. Increasingly reclusive, he shunned both lab and lecture, making appearances in the former at night and on weekends when the lab was all but empty.

Choosing to channel most of his energy into a national medical school association's needle exchange program in Cuba, for Reuven school became a second-tier priority as the semester crawled toward its conclusion. With his interest in anatomy reduced to reviewing Netter's *Atlas of Human Anatomy*, Reuven debated the propriety of memorializing an experience he found repugnant. Finally, shortly before the final exam, he contacted Vasan and requested that he be removed from the roster of memorial service speakers, explaining, "I just don't feel anything for it anymore."

Correspondent to Reuven's waning enthusiasm, Vasan's connection to the students, the cadavers and the process intensified as, each afternoon, the tangible results of the donors' benefaction were revealed in wide-eyed epiphanies destined to endure for the duration of 181 fledgling medical careers. Of all involved in disseminating the bodies of knowledge within Room 527B, no one had more gratitude than Nagaswami Vasan for the lasting impact of the gift resting upon each of the dissection tables.

Not until Vasan stepped up to the lecture hall podium at the start of the memorial service did the Class of 2002 realize the depth of that gratitude. With emotion exacerbating his thick Indian accent, Vasan leaned into the microphone to provide the epigraph of why they were gathered:

"You have probably, in your lives, given various gifts and donations. Well, these people donated their bodies in the hope that this donation of themselves will help you understand the human body and, in turn, help others who are suffering as you practice the art of medicine. These are very special people, they are people that you know in life, they come from all walks of life: They were parents and grandparents, husbands and wives. We don't know who they are and they don't know who you are and yet they donated their bodies to you.

"We don't know what their lives were like, we don't know their likes and dislikes or what their hobbies were. But we know all about them, we know their bones and muscles. Sometimes we can guess what went wrong with their bodies because we saw where things broke down. What they taught us, we will carry throughout our lives."

Six students followed Vasan to the podium.

Jennifer Heimall revealed how self-preservation had brought her to lab prepared to treat the cadaver as a working model, a strategy that dissolved with her emotional response to the dissection of the heart. After that, even as she and her lab partners decimated the body—Jennifer's table, number 28, had performed the full pelvic hemisection so envied by Table 26—the initial detachment evolved into a solace rooted in a sense of rebirth. The cadaver, she said, had "started a new existence in lab.

"I began to appreciate the beauty of the functional simplicity of the structures within him. In retrospect, I now know I learned from him the arrangements of nerves, organs and muscles but, beyond that, I learned much more. For starters, I learned that I really didn't enjoy cutting, so surgery is probably out as a career choice. Additionally, he led to jokes and closeness between me and my classmates that may not have happened had we not been going through the same common experiences together every day in lab. And finally, he taught me about the gift of human generosity in his ultimate gift of himself, to our medical knowledge."

A lot of people were surprised when Dan Mundy walked to the front of the lecture hall. The tallest member of the Class of 2002, Dan was a convivial enigma, known more for clowning around than sober reflection. Standing well over six feet tall, he frustrated Jen by his total uninterest in basketball; Jen would have loved to have had that height in the middle of ATP/2002's lineup. Equally confounding, Dan was among the handful of medical students who gathered in the courtyard for a cigarette break. Recalling the antismoking propaganda foisted upon them since grade school, his tobacco-free classmates couldn't fathom how anyone their age, let alone a medical student, could allow himself to become addicted to nicotine.

Attuned to the strengths and weaknesses of each student, Vasan spotted something within Dan that the course coordinator knew to be compatible with the spirit and tone of the memorial service. Dan Mundy didn't let him down:

"Other classes may quickly come and go during my stay in med school, but I'll never forget the first day of GA. It seemed every student in our class radiated excitement and anticipation. But what made this day so different, specifically, is that I felt for the first time that I resembled a doctor. Adorned in scrubs and long lab coat, I saw brief glimpses of my future, glimpses that became the driving forces that keep me going.

"There was, however, a deeper feeling of privilege, of initiation. I was about to begin a process sacred to the medical profession. In a way, it was my baptism into medicine, and I didn't completely understand the significance of the journey I was about to embark upon.

"Despite the countless hours spent in lab, the first month seemed to go by as fast as the first day. Inch by inch, we explored the complexity that lies beneath our skin. Daunted by the vast ocean of information I had to learn and anesthetized by the amount of time I had to spend in lab, some of the wonder of the first day disappeared. Our daily dissection acquired a tinge of monotony. And then came the day when we had to raise the shroud, exposing our cadaver's face.

"I'll never forget that day. My lab partners may not have known it, but I was scared. Just when I felt I was at ease with the activities of the GA lab, I realized something was amiss, my comfort far from complete.

"Countless times we'd been warned by our mentors not to detach ourselves from our patients. True, a certain amount of distance between a doctor and a patient is necessary. But when a doctor views a patient as an automaton or a textbook case, he's not fulfilling his obligations as a physician.

"Up to that point, I was never fully aware of our cadaver's humanity and, in a way, I was making the mistake we were warned not to make. As I stood before the concealed face, I could feel my heart beating faster. What were we about to witness? I've seen the faces of deceased loved ones before, but it was always in the warm, comforting atmosphere of a church or funeral parlor.

"I thought to myself, 'How can the deceased have dignity in this insensitive, barren laboratory, filled with cold, steel coffins?' Somehow, I found the strength to lift the shroud off our person's head. The gentleman's face seemed completely at peace, dignified and serene.

"I saw the face of a man who has given me the greatest gift a medical student can receive—the knowledge needed to heal. It is such a selfless act: giving the body he no longer needed so I may learn to heal the living who walk among us.

"I will always remember this gentleman's face, this man who gave to me so that I may give to others, an inspiration beyond words that I will never forget."

Be it on the basketball court or the gross anatomy lab, Jen preferred to let her actions do the talking. The downside of her high school athletic career hadn't been the hours spent on the practice court or field, but in the acclamation that came with success. Six years after she'd departed Little Egg Harbor, she remained dismayed at the plaque-laden wall of fame lining a hallway in her parents' home; the scrapbooks so lovingly assembled by her mother continued to be a source of embarrassment.

Given her aversion to the spotlight, Jen's first reaction to Vasan's entreaty that she speak at the memorial service was a stunned silence, followed by ready acceptance. Declining the invitation was not an option; being asked was an honor, an acknowledgment that Jen, the outsider who barely forty-eight months before thought her life would be spent dispensing medicine, not prescribing it, had worked her way into the inner circle.

Still, the thought of standing before a room of people gave Jen pause. As fond as she was of the cadaver, she would have felt much more comfortable had the topic focused on a clinical anatomical function rather than the state of her own emotions. Fortunately, med school allotted minimal time for excessive dwelling on any antipathy she had for public speaking. Besides, from the moment Vasan had asked for her participation, Jen knew what she wanted to say. On the night before the ceremony, dead tired from that day's exam and overcoming guilt for not studying physiology, she sat down to outline how she would express it:

"First, I'd like to say I am very grateful to have shared this experience with three wonderful people, Sherry, Udele and Ivan. They have made these demanding weeks not just bearable but enjoyable. I think you will all be physicians of the highest quality and I feel honored to have worked with you.

"These past few weeks have been very unique because, in death, we were able to find life. We were able to construct a vision about how each piece worked together to give this man's body life. We also developed a great admiration for this man, who we actually came to know very intimately. On reflecting on how his life might have been, Sherry quoted a verse from a song by Lucinda Williams, 'Did an angel whisper in your ear and hold you close, and take away your fears, in those long last moments?'

"For all he has given us in death, we just hope that he was rewarded in life. To conclude, I'd just like to share an experience I had over the past few weeks:

"It was one of those rare occasions when I actually got the chance to sit down and watch television, and I was watching a show called *Friends* that, um, I think a lot of you are familiar with. And two of the characters were debating the idea of a selfless act.

"One of the characters said there is no such thing as a selfless act, that any good deed that we might possibly do in the end will actually benefit us. So, through the course of the show, the one character's goal was to find the selfless act, to do everything she could to do something that didn't benefit her. But in the end, the other character was always able to justify why, indeed, it was selfish.

"And I kind of got caught up in all of this for the whole thirty minutes of, you know, trying to search for something that truly would be a selfless act. I mean, there had to be a selfless act out there somewhere. But at the end of thirty minutes, I couldn't come up with anything.

"About a week later, I was brushing my teeth one morning when all of a sudden it hit me that what these people did for us was, truly, a selfless act. And when we go back in there, into lab, I hope we'll always remember this generosity and never forget our first patient."

When all the words of tribute had been spoken, Vasan stood. Bowing his head, he asked the Class of 2002 to join him in closing the service with a minute of silent reflection, a quiet broken, presciently, by the muffled thud of helicopter blades announcing the arrival of another chopper on the University Hospital landing pad. The contemplative mien continued to resonate as the class repaired to the cramped rest rooms to change from street clothes into scrubs and lab coats for the rite designated as Checkout Day.

At 12:54 P.M., Udele for the last time opened the handles on the stainless-steel cabinet with the number 26, generated from a computer printout, taped at each end. Once the cabinet doors were secured under the table, Jen untied the tag that, in the previous day's exam, had asked the class to identify the artery to which the question was attached as well as the branch of the artery that "anastomoses [connects] with the branches of the iliac artery." Those answering correctly had written: "inferior mesenteric artery" and "superior rectal artery."

"You know, they used our body for every exam," Sherry pointed out.

"I know, isn't it great? I'm so proud of that," Jen replied.

"Yeah, but with all the great muscles he had, they still wound up tagging an artery," said Ivan.

"Pretty ironic," Sherry agreed.

As the remainder of the class meandered into lab, the remnants of solemnity left from the memorial service dissipated into muted satisfaction that outwardly demonstrated itself in congratulatory handshakes and platonic hugs. Camera in hand, one student attempted, without much success, to cluster her classmates for a group photograph. Once an anatomy class tradition, recording the scene for posterity fell into disfavor with the increased emphasis on maintaining the dignity of the cadavers. Eventually, the student convinced a handful of classmates to stand before the camera lens, but only after they'd been assured that no bodies would be included in the frame.

Paramount among the housekeeping chores on Checkout Day was placing the remains of the cadavers in the cardboard coffins to be transported to a crematorium in Linden, New Jersey. Content to leave the heavy lifting to her lab partners, Leslie Pooser circulated Lab C, saying good-bye as though it were the last time the Class of 2002 would be together. In a way, for Leslie, that was true. The recipient of special dispensation from the school, she wasn't required to take physiology until the next academic year, meaning that in many ways her classmates would move on without her.

Of everyone in the class, Leslie had undergone the greatest personal and intellectual metamorphosis. The scared, intimidated and naive Leslie Pooser, the one who spent that first afternoon praying for forgiveness, had transformed herself by Checkout Day into an emboldened medical student brimming with confidence.

Working at a table with Cary Idler and Persephone Jones, one surgically proficient before he'd taken so much as a single day of anatomy, the other patently compassionate, had helped Leslie's transfiguration immensely. So, too, had the studying alliance she'd formed with Sherry, the person in class closest to her age.

Intellectual and emotional development notwithstanding, the experience still occasionally weighed on Leslie. "Except for the ear [which would be studied more closely during neurology], we've looked at every inch, every inch of that cadaver. That's really disgusting when you think about it," Leslie said. The fact of the matter being that she thought about it little, having become so inured to what was going on that, as the semester drew to a conclusion, she couldn't remember the last time she'd come home to share the day's activities with Omari.

In the end, affirmation proved the ultimate reward for Leslie Pooser's struggle: After years of self-doubt, of scrimping and saving, of working through July and August while her fellow high school teachers took the summer off, gross anatomy had made all of it worthwhile.

In fact, the only decision changed by Leslie's tenure in lab was one made prior to med school that one day she would do unto others as someone had done unto her: "I will not donate my body to science. There's no way my fat ass will ever be on that gurney with a bunch of medical students digging through the fat and sticking a finger up my rectum looking for the internal pudendal artery," she said.

The lone covenant handed the students about placing the bodies in the temporary cardboard containers was that they make a clear notation, with an ink marker, of both the cadaver's table number and the end of the box in which the head rested, the latter an act of deference to religions that stipulate a corpse must be cremated headfirst. Other considerations, such as whether to empty the contents of pink plastic bags filled with tissues, organs and bones or place the filled bags inside the casket, were left up to the respective tables.

With Sherry off to the supply room to return the bone box, Ivan pulled a coffin off the pile stacked on the counter behind Table 30. The moment he set it on the floor next to their table, Udele and Jen joined him in placing into each of the box's corners the four plastic bags containing all of Number 3426's vital organs, skin and extraneous tissue. As the man had come to them whole, Table 26 decided so should he depart.

Without a word passing among them, Ivan then took a position at the head, Udele the torso and Jen the legs. Jen nodded and together they lifted Number 3426 over the side of the table and lowered him gently into the container. Afterward they paused, heads bowed reverently, appraising their first patient.

"Doesn't he look weird in the box?" Jen whispered huskily. "He looks different, like a body again."

Udele cleared her throat.

"He looks more like a dead person here than he did up there," Jen continued.

"I know. I think it's because now we can see all of him," said Udele.

Back from the supply room, Sherry joined the vigil. "We had a nice cadaver. A nice man," she said.

"Yeah," said Jen wistfully, picking a skin particle off the table, absently tossing it into the casket.

Ivan shrugged toward the clock. "We better get going," he said meekly. Already they were fifteen minutes late for a mandatory physiology lecture.

"Guess we'd better," said Sherry.

Ivan knelt on the floor. With a marker he wrote "26" boldly on each side of the box; at the front, he printed "26-HEAD." Ivan closed the left flap and began to fold it over the right when something caused him to hesitate.

"Bye-bye," Ivan Gonzalez murmured, pinching the flaps of the box shut, turning to receive, from Udele Tagoe, the length of masking tape that would seal it.

EPILOGUE

———————•———————

For nearly three months, the cardboard caskets remained stacked four-high on the counters where the Class of 2002 had deposited them before the mad dash to the physiology lecture. In May, shortly after the students departed for the summer, Roger Faison, Paolo Varricchio and David Abkin again fanned through the lab. The craniotomy a happily distant memory, no protective gear was required for the occasion of matching the paperwork from Faison's files with the numerical designations on the boxes. After making a match, Faison, Varricchio and Abkin carefully wrote the donor's surname on the appropriate container.

Encountering the casket upon which Ivan Gonzalez had affixed the number 26, Varricchio added the name Lewis.

It took some orchestration but finally, in early July, Faison managed to find a day when the Funeral Service of New Jersey could set aside a morning to pick up and transport the forty-five dissected bodies to the Rose Hill Crematorium in Linden.

Retracing the same route along South Orange Avenue taken by so many of their station wagons, the three-man Funeral Service team dipped into the loading dock tunnel beneath the medical school and hospital at 8:45 A.M. Two members of the crew headed for Faison's basement domain; the third stayed behind with the black Ryder rental truck to open two cans of Savarin regular-blend coffee, which he placed in the vehicle's cargo hold.

"An old New York City cop trick," Faison explained. "Supposedly, it gets rid of the smell. Whether it works or not, I really can't say."

Assisted in the lab by Faison, Varricchio and Abkin, the Funeral Service crew stacked the containers three to a gurney, which they wheeled through the supply room, down the elevator to the basement and through the double doors to the loading dock. Within an hour, the truck held forty-five cardboard caskets and two cans of Savarin.

As the final gurney passed through the morgue, Faison handed the crew chief forty-five folders with a yellow State of New Jersey Burial and Cremation Transit permit paper-clipped to the front of each one.

In Linden, the containers were wheeled to an open, flat freight elevator that descended down one floor to the massive refrigeration unit that would stave off further decomposition until the time when, in groups of three, they were brought back upstairs to the trio of Ener-Tek II Cremation Systems crematoriums operated by Rose Hill Cemetery. "It's the kind of job where what you see here is best not taken home with you," said the operator of one of the incinerators as it muffled ominously, giving off surprisingly little heat, nearby.

A month later, the cremains of the men and women dissected by the Class of 2002 arrived at the New Jersey Medical School. Pulling out her files, Essie Feldman went to work to ensure that the next of kin or designated caretakers received the 8 x 6½ x 4½-inch cardboard boxes in an expeditious manner.

In mid-August, Connie Lewis received in the mail a package from NJMS.

Since her husband's death nearly two and a half years before, there'd been some discussion among the family about how to dispose of his ashes. As a veteran, Tom was entitled to burial in a national cemetery, an interment favored by Connie. Remembering their father's wishes, a few of the children pushed to have the ashes spread upon the waters where he spent the last years of his life crabbing with the grandchildren, Barnegat Bay.

Summer dragged into fall without resolution and, over the winter, Connie departed for Florida, leaving the cardboard box in the care of Chris and his wife, Susan. When their mother returned home the following spring, the friendly debate renewed itself, dragging into yet another summer until, finally, Connie prevailed.

On a sweltering Sunday late in the first July of the new millennium, the family of Thomas Lewis, Jr., gathered at Arlington National Cemetery, where, with full military honors, they laid to rest the man they knew and loved as husband, father and grandfather.

A month later, Udele Tagoe, Sherry Ikalowych, Ivan Gonzalez and Jennifer Hannum began their third year of medical school.

ACKNOWLEDGMENTS

This book would not exist were it not for five individuals who each played a significant role in bringing it to the fore. First among equals, Noah "Kenny" Murray pitched the idea that we spend a semester chronicling the lives of medical students in an anatomy lab. I'd be remiss by not admitting that my first reaction was far from enthusiastic. But Kenny, as is his way, kept hammering away until finally, as much to shut him up as anything else, I contacted the University of Medicine and Dentistry–New Jersey Medical School. The resulting newspaper story, highlighted by Kenny's award-winning photography, led to the publication of this book.

Long before I ever believed I could successfully undertake a project such as this, Kevin Coyne demonstrated unwavering faith and confidence in me. When it finally happened, Kevin was there each step of the way, providing inspiration, insight and a level of friendship beyond value.

Of everyone, Lisa Kevis suffered most the insecurity and self-doubt that are the bane of writers, or at least this one. Her love, prodding and a heretofore untapped expertise in editing means that this achievement is Lisa's as well.

I owe a major debt of gratitude to my agent, Elly Sidel, for her confidence, support, unending encouragement and for making all this possible.

Lisa Drew honored me by providing the opportunity to write for one of the most highly regarded editors in publishing. Thanks also to Jake Klisivitch.

As a journalist, I've always maintained that a writer is only as good as his or her primary sources, and in Udele Tagoe, Jennifer Hannum, Ivan Gonzalez and Sherry Ikalowych, I had four of the best. For fourteen weeks they tolerated my overhearing their conversations, prying into their personal lives and interrupting their education. I would, without hesitation, entrust the care of my son to any one of these fine and astute physicians.

Dr. John H. Siegel, professor and chairman of the Department of Anatomy, Cell Biology and Injury Science, has not once but twice granted me access to the gross anatomy lab at the New Jersey Medical School while sharing with me his infinite wisdom about a most complicated subject. Along the same lines, every man or woman, no matter their profession, should be blessed in their education by the erudition and compassion Dr. Nagaswami Vasan conveys to each of his students.

Many members of the NJMS staff, faculty and administrators contributed to this book, especially Roger Faison, Dr. Anthony Boccabella, Dr. Paolo Varricchio, Dr. George Heinrich, Dr. David Abkin, Dr. David DeFaux, Dr. Zolton Spolarics, Dr. Michael Rose, Dr. Elizabeth Alger, Dr. Joan Limon, Mary Ford-Mathis, Dr. Shamshad Gilani, Dr. Arvindkumar Suthar, Dr. Barry Grunwerg, Dr. Ajit Dhawan, Dr. Mark Nathanson, Dr. Christopher Leung and, of course, Essie Feldman.

Other medical sources included Dr. Jack Coulehan and Catherine Belling of the State University of New York–Stony Brook, Dr. Rita Charon of the Columbia University College of Physicians and Surgeons, Dr. Sandra Bertman of the University of Massachusetts Medical School, Dr. Todd Olsen of the Albert Einstein College of Medicine and Dr. Peter Dans of Johns Hopkins University.

Whether they were aware of it or not, many members of the Class of 2002 helped me in large and small ways during my time at NJMS. They include, first and foremost, Leslie Pooser, Christine Ortiz, Leah Schreiber, David Murphy, Jennifer Heimall, Tamanna Patel, Elizabeth Garcia, Bradd Millian, Matthew Hosler and Priya Singh as well as Annette Pham, Matthew Gerwitz, Cary Idler, Janet Chen, Persephone Jones, Alex Flaxman, Kevin DeBrancanga, Rachel Altman, Madhukar Bhoomifeddi, Reuven Bromberg, Tahl Colen, Shawn London, Mohan Madhusudanan, Stevan Marjanovic, Angela Huang, Mark Jablonski, Edward Kim, Marcia Klein, Michael Picone, Joyce Prophete, Rajesh Raman, Gayatri Rao, Nishant Shah, Daniel Mundy, Ramon Nunez, Odujebe "Dapo" Oladapo and Avi Deener. Thanks, also, to David Palencia and his classmates at the Albert Einstein College of Medicine.

By sharing the details of his extraordinary life, the family of "Thomas Lewis" helped put into perspective the magnitude of the gift he provided all of the Class of 2002 and especially Jen, Udele, Ivan and Sherry. For that, and their kindness to me, I am extremely grateful.

I need also give my appreciation to Dr. David Gutierrez, Dr. Theresa Auld, Dr. Alan Nasar and Dr. Howard Frauwirth—the four physicians who were the subjects of the newspaper story that resulted in my receiving a contract to write this book. The editors and graphic designers who contributed to that story also receive my thanks: Joe Green, Wally Patrick, Bob Williams, Sanne Young, Ray Olwerther, Jody Calendar, Andy Prendimano and Harris Siegel.

Thanks also to Brenda Barbosa, for her bilingual assistance; Ruth Meaney and Andie Meaney, for adjusting their schedules; Kathleen Lanini, Father Hank Hilton and Sarah Hilton; Jane Herman; Emily Krauss and Maris Krauss; and the families of Udele, Ivan, Sherry and Jen.

In the career of every journalist there are editors who not only edit but teach. Thus I would be remiss if I didn't acknowledge that each and every story I write carries within it lessons taught me by the immeasurable talents of Arlene Schneider, Virginia Wray Richards, Dennis Hartig, Nancy Moore, Jim Mann, and the late Robert E. L. Baker.

Finally, in the writing of this book, as with all I do, Bix was with me in spirit, and sometimes in person, during the composition of every page.

INDEX

Index